Embedded System Design on a Shoestring

Achieving High Performance with a Limited Budget

Embedded System Design on a Shoestring

Achieving High Performance with a Limited Budget

by Lewin A.R.W. Edwards

 Newnes

Amsterdam Boston Heidelberg London New York Oxford
Paris San Diego San Francisco Singapore Sydney Tokyo

Newnes is an imprint of Elsevier Science.

Copyright © 2003, Elsevier Science (USA). All rights reserved.

 Recognizing the importance of preserving what has been written, Elsevier Science prints its books on acid-free paper whenever possible.

Library of Congress Cataloging-in-Publication Data

ISBN: 0-7506-7609-4

British Library Cataloguing-in-Publication Data
A catalogue record for this book is available from the British Library.

The publisher offers special discounts on bulk orders of this book.
For information, please contact:

Manager of Special Sales
Elsevier Science
200 Wheeler Road
Burlington, MA 01803
Tel: 781-313-4700
Fax: 781-313-4882

For information on all Newnes publications available, contact our World Wide Web home page at: http://www.newnespress.com

Contents

Acknowledgments iv

Chapter 1: Introduction 1

Chapter 2: Before You Start—Fundamental Decisions 9

General Microcontroller Selection Considerations 9

Choosing the Right Core 13

Building Custom Peripherals with FPGAs 19

Whose Development Hardware to Use—Chicken or Egg? 21

Our Hardware Choice—The Atmel EB40 29

Recommended Laboratory Equipment 30

Free Development Toolchains 32

Free Embedded Operating Systems 36

GNU and You—How Using "Free" Software
Affects Your Product 44

Choices of Development Operating System 51

Special PCB Layout and Initial Bring-Up Rules
for the Shoestring Prototype 53

Hints for Surface-Mounting by Hand 62

Choosing PCB Layout Software 65

Chapter 3: The GNU Toolchain 71

Building the Toolchain 71

Overview of the GNU Build Environment 76

GNU Make and an Introduction to Makefiles 80

Contents

Gas—The GNU Assembler 87
    Comments 88
    Symbols and Labels 88
    Code Sections and Section Directives 90
    Pseudo-Operations 96
    Conditional Assembly Directives 108
    Macros, Assembler Loops and Synthetic Instructions 111
Ld—GNU Linker 114
    Introduction 114
    The SECTIONS command 118
    Symbol Assignments, Expressions and Functions 119
    Output Section Descriptions 124
    Overlay Section Descriptions 127
    Emitting Data Directly into the Executable 131
    Input Section Descriptions 132
    Named Memory Regions 134
    Special Considerations for C++ 136
    Further ld Information 137
Converting Files with Objcopy 138
Objdump—Check Your Executable's Layout 139
Size—Check the Load Size of Your Executable 143
Gdb—The GNU Debugger 143
    Invoking and Quitting gdb and Loading Your Program 145
    Examining Target Memory 148
    Breakpoints and Other Conditional Breaks 149
    Getting Further Help 151

Chapter 4: *Example Firmware Walkthroughs
and Debugging Techniques* *153*

A Quick Introduction to ARM and the Atmel EB40 153
First Step—the LED Flasher (in Assembler) 158

Bringing Up a Simple C Program—
The LED Flasher (in C) 167

Writing a Simple Flash-Loader
(and Inspecting Memory with gdb) 172

A Simple ROM-Startup Program 180

A Complete ROM-Startup Application in C 185

Blind-Debugging Your Program 194

Miscellaneous Glue—Handling Hardware Exceptions
in C with gcc 199

Chapter 5: *Portability and Reliability Considerations* 203

Chapter 6: *Useful Vendors and Other Web Resources* 221

Index of CD-ROM Contents 223

About the Author 227

Index 229

Acknowledgments

The author would like to extend his sincere thanks to the following individuals and corporations who have contributed directly and indirectly to the publication of this book:

- Atmel developer support
- Cadsoft Computer, Inc.
- Cirrus Logic developer support
- Michael Barr
- Don McKenzie of dontronics.com
- Spehro Pefhany
- Rob Severson of USBmicro
- Sharp Microelectronics developer support

In keeping with the open-source nature of this book's subject matter, the manuscript of this work was developed entirely using the free open-source OpenOffice.org office productivity suite, under Red Hat Linux 8.0.

## Chapter 1

Introduction

There exists a large body of literature focused on teaching both general embedded systems principles and design techniques, and tips and tricks for specific microcontrollers. The majority of this literature is targeted at small 8-bit microcontrollers such as the Microchip PIC, Atmel AVR and the venerable 8051, principally because these devices are inexpensive and readily available in small quantity, and development hardware is available from a variety of sources at affordable prices. Historically, higher-performance 16- and 32-bit parts have been hard to obtain in small quantities, their development toolchains have been prohibitively expensive, and the devices themselves have been difficult to design around, with tight electrical and timing requirements on external circuitry necessitating very careful hardware design. A dearth of royalty-free, open-source operating system and library code for these processors also meant that developing a new project was a huge from-the-ground-up effort.

However, over the past few years we have simultaneously seen the size and price of 16- and 32-bit cores fall, and the development of many highly integrated parts that enable the easy development of almost single-chip 32-bit systems. In addition, many readily available appliances now contain a well-documented 32-bit microcontroller with ample flash memory, RAM and a variety of useful peripherals such as color LCDs, network interfaces and so forth, which can be exploited by the cunning embedded developer as a ready-made hardware platform. Cross-platform assemblers, high-level language compilers and debugging tools are available free for the downloading and will run satisfactorily on the average desktop PC; it is no longer necessary to spend tens of thousands of dollars on proprietary compilers and special workstations on which to run them.

As these systems have increased in complexity, to a certain extent the degree of specialization required to develop them has decreased. This might sound paradoxical, but consider the fact that high-end 32-bit embedded systems, and the tools used to develop for them, are effectively converging with the low-end mainstream PC. The skills required to develop an application for embedded Linux, NetBSD or Windows CE are by intention not radically different from the skills used in developing applications for the desktop equivalents of these operating systems (though of course different coding best practices usually apply in embedded environments). In most cases there are mature off-the-shelf operating systems available ready-to-run for the common hardware reference designs and manufacturer-supplied evaluation boards, so we are usually spared even the initial bring-up phase and much of the effort required to debug device drivers.

Given a working hardware platform with reasonably well-documented components, the only task for which traditional embedded expertise is absolutely necessary is to create the necessary bootstrap and "glue" code to get a C run-time working on the target platform, and perhaps create drivers for some peripherals (and as discussed above, even this step can often be skipped if you are building around a reference platform). From that point on, most of the programming work to be done runs in the application layer and can be accomplished using high-level languages. There is a large workforce available almost ready-trained for this type of coding.

The end result of this evolutionary process is that it is now well within the financial and logistical reach of a small company or even an individual hobbyist or student to develop (or at least repurpose) advanced embedded systems with exciting functionality provided by these high-performance parts. Unfortunately, however, device vendors' support infrastructures are still geared towards large-scale commercial developers. This raises two major obstacles:

1. Development hardware for high-end parts is, by and large, still too expensive for the average hobbyist or student. This is partly a chicken-and-egg problem; the only source for evaluation boards for 32-bit parts is usually the chip vendor, because there isn't sufficient third-party interest to see third-party evaluation platforms developed. The resulting small volumes and lack of competition conspire to keep prices high. From hearsay, it seems likely that some chip vendors also have an intentional policy of excluding small customers from purchasing high-end devices. In many cases, the evaluation board is unavoidably expensive because it is designed to showcase what can be achieved in a maximally configured appliance in the chip's target market; consequently, the board has a large number of peripherals.

2. In order to ensure continued support from major embedded toolchain vendors, chip designers usually recommend only specific development environments, all of which are extremely expensive. It is still quite rare to find explicit manufacturer support for freely available compilers and debuggers, despite the widespread adoption of such tools in the industry at large. One underlying business reason for this is that in order for a new part to be credible, it should be supported from its release by well-known commercial toolchain vendors. To encourage active interest from the developers of these toolchains, and to reduce their own support workload, the chip vendors generally avoid mention of free, user-supported tools.

The main object of this book is to illustrate some methods of overcoming these obstacles and realizing exciting projects around today's high-performance chips. A strong secondary objective is to assist developers in migrating from the coddled environment of one-click graphical integrated development environments to the command-line tools typical of free toolchains. Although there are many references for this available on the Internet and in printed form, the authors of such guides usually do not archive the tool versions they discuss in their text, and their instructions often contain information that is not applicable to the currently available versions of the tools. It can therefore be difficult for the neophyte to know which of his or her problems are genuine coding errors, and which are simply the result of documentation inconsistencies.

Given these goals, this volume is aimed at the following groups of readers:

Hobbyists and students. These people are typically financing the acquisition of development hardware and software out of their own pocket. They do not necessarily expect a direct cash return on this investment, and they are usually working alone or in small groups. This type of reader is interested in solutions that involve minimal startup expenses and don't require large engineering teams.

Entrepreneurs. This class of reader has a product idea that needs to be at least prototyped so that it can be shown to potential investors with the aim of securing development financing. Not only does a real prototype have a much better "wow" value (and hence a better chance of attracting investors) than a sketch and verbal description, but developing the prototype will reveal and perhaps solve many of the engineering problems to be encountered in making the real product. This obviously translates directly into a shorter time-to-market.

Engineers working alone or at small companies. Small engineering houses that currently work with 8-bit systems may realize significant gains by moving to 32-bit parts. In many cases, functionality provided by dedicated hardware in the 8-bit system can be synthesized in firmware on the 32-bit system – this has obvious profit benefits because the company can thin its inventory to just a few standard hardware platforms, differentiating products by means of firmware features. Customer satisfaction can also be enhanced, since new features can be added with simple firmware upgrades. Because the hardware is standardized, reference designs have a longer lifespan, meaning that new projects may only require incremental software changes – potentially an enormous saving in development time. However, the leap from 8 to 32 bits is a significant one and it can seem prohibitively expensive in the short term even when the long-term benefits are well understood. After reading this book, it should be clear that the up-front investment is not necessarily huge, and it may be time to make the big jump to 32-bit cores.

Engineers working on pet projects at large corporations. Although such corporations have the wherewithal to fund a traditional development process, it can be hard for an R&D engineer to inveigle management into bankrolling projects that don't have clear delivery dates. Developing an exciting demonstration of the project and presenting it to management is an excellent way of drawing official attention to an idea, escalating it from a pet project to an official project. Many corporations have a cash bonus program in place for employees who develop new ideas like this, so there is a clear personal benefit to the engineer, as well as a new profit opportunity for the employer.

Small to medium-sized corporations trying to reduce their reliance on outsourcing and increase their profit margin by developing custom solutions in-house rather than buying or contracting turnkey solutions from external vendors.

In this book we will be working on fragments of a completely fictitious project. The actual functionality of this device isn't important the value of this book lies in the tools and techniques discussed, and having a project to work on simply allows me to show concrete examples rather than talking entirely in generalities. For reasons described later, I have chosen an ARM-based microcontroller for my example project, but no ARM-specific experience is required to understand the concepts presented here. By the end of this volume, among other things, you will understand how to get the free GNU toolchain built for a specific target (ARM is illustrated, but the steps are identical for other targets), and you will have a good introduction on how to use the various components of this toolchain, with specific emphasis on functionality of interest to the embedded developer. Such functionality is often glossed over in general discussions of the GNU toolchain, and it can be hard to infer the mode of operation of these tools simply by inspecting example code.

Please note that this book is not an introduction to embedded systems per se; it is intended to help a reasonably experienced developer identify and use a variety of inexpensive or free tools and other resources in lieu of costly commercial alternatives. Throughout the text, I am assuming mostly that the reader has a basic understanding of the C language and experience with some level of assembly language programming. I have also assumed at least minimal experience with command-line UNIX (simple directory listing and manipulation commands such as ls, cp, mkdir, rm only). The specific aims of this text are:

- To describe design and component selection rules specific to the engineer with a severely constrained budget and no significant relationship with IC vendors.

- To describe techniques for PCB layout and assembly that will enable the reader to build complex 32-bit systems using hobbyist-grade laboratory equipment.

- To provide basic documentation on building and using the GNU toolchain, particularly relevant to programmers with experience on other toolchains who need to understand the syntactic idiosyncrasies of the GNU tools.

- To provide some simple examples illustrating how to use GNU tools to perform the most common "up and running" tasks required to bootstrap an embedded system.

- To provide guidelines on best practices to employ when developing demonstration products on general-purpose hardware, with the intention of later developing real, marketable hardware.

In the sections that discuss hardware and laboratory tools, I assume some experience with the hardware side of embedded systems design. Depending on what type of project you're attempting, and how you approach it (for example, you might choose to build your device around a commercially-available single-board computer, rather than designing your own hardware), not all of this information will be relevant to your case.

You will see that throughout this book, I will mention specific products and in some cases specific prices. This text is, however, not a catalog; all prices (though correct at the time of writing) are mentioned purely to give you a feeling for what kind of money you will need to invest in projects of the kind we are discussing. Furthermore, I feel it necessary to point out that I have not received any consideration from the suppliers of any of the software or hardware components I discuss here (except in some cases, permission to put demonstration versions of their software on the accompanying CD-ROM). Part of the realism of this book is, intentionally, that all the equipment and tools that I mention were acquired with personal funds out of my "hobby" budget. Readers should also note that products mentioned here are somewhat US-centric, since I am currently located in the United States.

Regardless of the nature of the project you attempt, or the path you follow to develop it, I hope this book will help to dispel the almost mythical aura surrounding high-end embedded system development. Readers are encouraged to visit my web site (www.zws.com), where I will post corrections and updates to this text. I can be reached via email as sysadm@zws.com; I can't guarantee to reply to every email I receive, but I will do my best. I can also usually be found in the Usenet group comp.arch.embedded, and I invite readers to join in the discussions there.

## Chapter 2

Before You Start—
Fundamental Decisions

General Microcontroller Selection Considerations

The start of a complex embedded project, particularly in a small organization without engineers who can be dedicated full-time to component procurement, can be extremely stressful. Until a first-round prototype is built and tested (and often even after this stage), it is usual for hardware requirements to be at least slightly vague, particularly vis-à-vis the exact breakdown of which functions are expected to be integrated into the microcontroller and which will be off-chip. As the design engineer, some of your goals are obviously ease of firmware and hardware development, low bill-of-materials cost, and reliability of sourcing. You will probably start with a list of hardware requirements, and match those up against selection matrices from different vendors to find a part that has as many of your features as possible on-chip.

At this point, what you really want is a vendor-neutral parametric search engine, where you can select the performance and peripherals you want, and obtain a list of suggestions collated from *everybody's* catalogs. Unfortunately, most of the search facilities available online leave much to be desired. Many manufacturers don't have full parametric search engines available, and those that do obviously only list their own parts. Third-party search engines do exist, but they are usually premium services for which you will have to pay—and again, they only list products from manufacturers with whom they have a relationship. Also, the total startup cost of development—evaluation boards, tools, etc.—is an important factor to us (for some readers, perhaps even more important than the unit cost of the microcontroller), and this will not be listed by parametric search engines. Finally, as with any other search facility, it can be difficult to match your needs with the list of keywords provided in the search engine.

This is one occasion where there is no substitute for peer support. Even if you think you've found a perfect match already, it's well worth searching Usenet archives (groups.google.com) for discussions on similar applications to your own. A carefully phrased question may lead to even more useful suggestions. Even if you are intimately familiar with every IC vendor that impinges on your industry, you might miss a new product announcement and thereby not know to check manufacturer X's catalog. Sometimes the only clue you need to lead you to the right part is the information that manufacturer X makes 32-bit microcontrollers! Furthermore, other engineers who have worked with the part may be able to point you to low-cost third-party evaluation platforms or off-the-shelf appliances that can be used as demo boards, and they will be better-positioned than anyone else to give you relatively unbiased opinions on real-world difficulties of using a specific device.

In the early days, it is also doubly hard to make an optimal price-performance choice, because the selection sheets generally won't show pricing. For any part that can't be bought anonymously off the shelf (and unfortunately the majority of 32-bit microcontrollers fall into this category), most chip vendors expect you to establish a relationship with their distributors. This can waste a lot of time in profitless face-to-face meetings. My own experiences with local reps and distributors in the United States have been *very* patchy, but I have often found that their knowledge of the 32-bit parts on their line card is limited to whatever bullet points the manufacturer printed on the sales literature. The distributors want accurate annual usage forecasts before they will give you sensible pricing, and they obviously have little or no incentive to deal with small-volume purchasers like students or hobbyists. Political difficulties related to sales commissions also arise when you are designing the product in one country, but intend to manufacture it in another. Furthermore, the distributors and reps will be most likely to quiz you on your other requirements and try vigorously to sell you other parts from their line card. Although this possibly has some marginal convenience benefits if you intend to source and manufacture locally, it certainly isn't the ideal way of minimizing the bill-of-materials cost of your product.

It's all too easy to become trapped in an endless circle trying to seek an optimal solution to all these problems, so you shouldn't attempt it. Recognize from the outset that this is a classic "traveling salesman" problem (perhaps even in the literal mathematical sense) and that your goal is merely to find an *acceptable* solution in time to finish your project and send it to the factory (or submit it to your professor, if you're a student). Your goal is not to find the best possible solution. If your team has enough personnel to dedicate a lot of person-hours to sourcing components, you will probably be able to find a better solution than the one-person "team" scouring catalogs on a time limit, but a suboptimal one-person solution can always be refined later if the project goes into production in quantities that justify it. As in any other industry, our goal is to develop a product that works properly and is ready to manufacture in a timely fashion.

With that said, I employ the following useful heuristics to filter my short list for 32-bit microcontroller selection:

- The device should be available for anonymous online or catalog ordering in single-piece quantity from at least one major distributor. (In the U.S., the big names commonly mentioned are Digi-Key, Newark and Avnet Marshall. Digi-Key and Newark in particular have very broad inventories and generally allow purchases in small quantity. Avnet Marshall seems to cater more to manufacturing rather than prototype runs; they typically have 25 or even 250-piece minimum orders on parts).

- Full datasheets for the device should be available without requiring a non-disclosure agreement or committing to any kind of purchase.

- A low-cost development board should be available for the part; either the manufacturer-recommended board, or a third-party board, or even some appliance based around the chip, as long as sufficient documentation exists to enable use of the appliance as a testbed for your own code. You should also ask the manufacturer and distributor if loaner boards are available; if you can borrow a board for a month or two, it will be enough to get at least bootstrap code up and running, and establish a basic level of familiarity with the microcontroller. You can then move to your own hardware and return the evaluation board.

- There should be a direct technical contact available at the chip vendor, at least for emergency issues; it should not be necessary to route all questions through distribution. (Note that I'm **not** advising you to abuse such a privilege—if you have a direct manufacturer contact, it's best to contact him or her only when absolutely necessary. But there are times when a complex problem will take weeks to solve when there are several layers in the communication chain, versus only a day or two if you can communicate directly with the cognoscenti at the chip manufacturer. As a small customer, the less you use this resource, the better chance you will have that your next urgent question will be answered speedily.)

- The device should have been shipping to OEMs for at least three to six months.

- The core should be supported by the GNU toolchain.

- There should be at least one currently shipping commercial product that uses the device, and the larger the market for this device, the better. All too often, parts that are consumed only by small niche markets are discontinued in favor of parts with more general applicability.

These are not absolutely binding rules (in particular, the last one can be hard to obey for a brand-new part) but they provide a good way of thinning a short list of any undesirable parts that are going to cause logistical problems later. The first criterion above is especially important to note, because it can give you some idea of the part's longevity. One little-mentioned fact of the microcontroller industry is that very few high-end parts are just designed for the marketplace in general; many of the "standard" 32-bit parts and ASSPs started life as proprietary ASICs developed under contract for some specific electronics manufacturer. These contracts typically have large guaranteed order quantities and forward-planned production schedules. However, once that manufacturer's exclusivity expires, the chip vendor is free to sell it to other people, if it conceivably has any generally applicable function.

The first step in this process is usually to offer the part quietly to other existing customers or to carefully selected others, without a highly visible product announcement or other publicity. This small group of privileged customers will, again, work on large volume pre-orders with long-term schedules. If a chip goes on from this into retail distribution channels (such as Digi-Key and other stores catering to small orders), it is a very good sign, because it usually means one of two things:

1. The chip vendor is seeking to gain market share in the field addressed by this part, and is pushing it heavily (also implying that excellent support will be available both from the manufacturer and other users), or

2. The product is so wildly successful that the chip vendor is producing reasonable quantities of it in advance of any firm order, in expectation of future unscheduled orders.

In either case, the part is in wide-scale production, and it is a fairly safe bet to design it into your product. You can be reasonably certain that the part will not be discontinued in the immediate future.

Choosing the Right Core

Unfortunately, even with the greatest care in choosing parts that appear to be supported for the long term, there are never any guarantees. Parts are discontinued or superseded all the time for marketing reasons that are sometimes not obvious and far from predictable. For that matter, sometimes your requirements change slightly and your previous choice of microcontroller is suddenly no longer suitable. This is particularly annoying when a design change of this sort is a result of entirely external forces. I have been involved in several projects where the microcontroller has been changed just before production, or even after production starts, simply because of sudden supply shortages of other parts.

Obviously, the more careful you are in choosing a part that *exactly* meets your requirements, the more disruptive it is likely to be to have to substitute a different part. A large customer might be able to guarantee the chip vendor enough volume for them to continue occasional production runs or even perhaps migrate an old part to a new process and continue general production. Since we're going to be a tiny customer, we won't have this luxury.

The only truly effective preparation for this inevitability is to anticipate it and pick a microcontroller based around a popular core to minimize the workload of porting to a new processor when circumstances demand it. Generally speaking, there are seven very widely used 32-bit cores on the market at the moment: Motorola 680x0, Intel x86, PowerPC, MIPS, SuperH, and ARM[1]. Numerous less popular or proprietary architectures also exist, of course; many of these are associated with specific applications such as laser printers or DVD players.

At the risk of antagonizing its userbase, I recommend against choosing the 680x0 series for a new design. Usage of this core appears to be in decline, and it is perhaps actually close to the end of its life; the principal consumer use at this time is in PalmOS® devices. These PDAs are now migrating towards ARM, and even Motorola has introduced an ARM-cored processor as its new flagship PDA part. The entry-level laser printer market, which formerly consumed a lot of MC68000 and MC68008 parts, has largely been dominated by cheap devices that lack a rasterizer (they rely on the driver software running on the attached PC); so they only require simple servo control on the printer mainboard.

Architectures based around the high-end x86 family (and code-compatible parts from AMD, National Semiconductor, Via Technologies, etc) have some immediate advantages:

■   You can use almost any PC-compatible operating system, and free software development tools.

---

[1]   Note that I am only mentioning general-purpose microprocessor cores here. DSPs are a separate world beyond the scope of this text.

- Installing operating systems is simple; in most cases there are automated installers that will probe your hardware combination and automatically install appropriate kernels, drivers etc. Compare this to the norm with embedded systems, where you will need to look at the board, work out the hardware configuration yourself, and sysgen the kernel and driver set on external hardware, probably using a cross-compiler.

- It is simple to interface literally thousands of peripheral components for almost any imaginable function. Because these components are produced for the consumer market, with its enormous volumes and bloodthirsty price competition, peripheral components are cheap and fairly easy to acquire.

- Driver support exists (within the framework of most off-the-shelf operating systems) for almost any piece of hardware you could want to attach to your system.

- Highly integrated mainboards are available with many possible combinations of peripherals, in a wide variety of form factors.

- Migrating to a slightly different hardware platform due to shortages of support parts or evolving customer needs is relatively simple; in many cases, it simply involves recompiling and reinstalling the operating system and preparing a new master disk image for duplication.

Having extolled the obvious virtues of these parts, I must also point out some of the downsides:

- x86 parts are very expensive, in production quantities[2], compared to RISC alternatives of comparable performance. This may affect your ability to commercialize your device.

- There are relatively few x86 variants that are true "system on chip" devices, so you are likely to need quite a bit of external hardware in addition to the microprocessor itself. Often, in order to obtain one specific function, you will need to add a

---

[2] This statement needs qualification. Although the x86 CPU is quite expensive, you may find that a given system configuration is cheaper when built around an x86 than a RISC processor such as PowerPC, because of the significant economies of scale in producing large volumes of the x86 board.

complex multifunction part because the single function you want isn't available as a discrete component. Again, this brings up your system complexity and total bill-of-materials cost.

- x86 has significant power consumption, heat and size disadvantages. (The Transmeta Crusoe x86-compatible device combats these disadvantages, but it is currently rather expensive and not very many vendors have products based around this microprocessor).

- Modern x86 parts and their support chips are very high-speed devices in dense packages. It is virtually impossible to hand-prototype your own design based around these parts; unless you want to spend many thousands of dollars on equipment, at the very least you will have to contract out some assembly work.

- PC peripheral ICs often have very short production lifespans; twelve to eighteen months is not uncommon, so ongoing sourcing may be an issue.

- Code to cold-boot a "bare" PC platform is usually very complicated, because you have to replace numerous layers—motherboard BIOS, expansion card BIOS, and various OS layers. The CPU architecture is also complex.

- Although I personally don't consider this to be a serious downside, it bears pointing out that JTAG-based or other hardware debugging systems aren't usually available on commercial single-board x86 computers.

To the people for whom I have written this book, I recommend x86 as the platform of choice if you are either building just a few of your appliance, or if you are prototyping something and want to pull together a lot of miscellaneous hardware features without spending a lot of time debugging the hardware design. It's also a good choice for an initial production run that you can ship to early adopters while you are developing a cheaper second-round customized hardware design. There are other special situations where you might find x86 to be a good choice, but these are the major ones.

Of course, you aren't restricted to using Intel parts; for instance, one x86-compatible part that is fairly popular in embedded applications is the Geode series from National Semiconductor (based on intellectual property acquired from Cyrix). This part was designed for Internet appliances and can be found in several such devices on the market today. There also exist numerous single-board computers built around Geode chips, with various peripheral functions according to the intended application. Geode was also used as the reference platform to develop and showcase the new Microsoft Smart Display device, so the product family is likely to be supported for quite a while.

Using x86 also doesn't mean that your device needs to have a large PC motherboard and expansion cards inside it. Unless your needs are highly specialized (and perhaps even if they are), it is probable that you will be able to find a single-board computer with most or all of your required hardware already integrated. These boards range in size from "biscuit PCs" with the same footprint as a 5.25″ disk drive down to a fairly new standard (consisting of a user-designed baseboard holding an off-the-shelf module containing the CPU and some peripherals) usually referred to as ETX. Embedded computer boards like this typically have PC/104 expansion buses (a condensed, stackable version of ISA using 100 mil headers) or Mini-PCI. Some of the larger boards will have regular PCI slots, but these start to make the overall system unavoidably rather bulky, approaching the size of a normal slim-line PC.

Note that PC-compatible SBC pricing falls into two widely separated categories: industrial and commercial. Industrial SBCs are *extremely* expensive; at least twice the cost of commercial versions. Commercial SBCs, though substantially more expensive than consumer grade PC hardware of the same nominal specifications, are a much better choice for the budget-constrained purchaser. Many SBC vendors specialize in industrial automation only, so if the prices you are being quoted seem unrealistically high, you should investigate other vendors before concluding that x86 is too expensive for your project.

Moving onto the RISC platforms, MIPS, SuperH and PowerPC are good candidates for many applications, and in particular the SuperH family is large and contains a wide variety of useful devices, though MIPS seems to be a more widely licensed core in third-party ASICs and ASSPs. PowerPC seems to be found mainly in applications requiring very high performance. In evaluating all of these parts for various projects, I have found them to be fairly difficult to develop with on a shoestring budget; evaluation hardware is usually costly, and most variants of these parts are not readily available to buyers who are unable to demonstrate a need for large quantities. However, all of these cores are likely to remain available and well-supported for the foreseeable future, so they are all viable choices as long as you can obtain development systems and parts.

At least in the case of SuperH and MIPS, your cheapest path to a prototype based on these parts is generally to repurpose some existing piece of hardware such as a PDA; for PowerPC, I would suggest buying a commercial single-board industrial control computer based around the chip of interest. Be warned that this is likely to be expensive; PowerPC boards don't have the same kind of mass-market pricing as x86-compatible boards and you can expect to pay between two and three times as much for a PowerPC SBC as for a comparable x86-based board.

Bearing the above discussion in mind, unless some of the Intel arguments apply to your case my primary recommendation for a 32-bit embedded platform is ARM. This architecture has many important advantages (some of these are also applicable to the other RISC platforms mentioned above, of course):

- It is a mature, well-understood architecture with a solid engineering history and many refinements. The large number of current licensees and now-shipping parts makes ARM a very safe bet for future availability.

- The cores are small and have excellent power consumption vs. performance characteristics.

- Many features—coprocessors, external bus widths, memory-management unit, cache size, etc.—are tunable by the chip designer, meaning that a core variant can be found to meet almost any performance/size/power requirement.

- There are a huge number of attractively priced standard, custom and semi-custom parts on the market with a wide variety of integrated peripherals.

- Since ARM provides reference designs for many different peripherals as well as the core itself, there are often similarities in peripheral control on different ARM implementations, even from different vendors. To take a trivial example, code to send data out of a serial port can usually be ported from one ARM variant to another with little effort.

- Partly because of the above factors, there is a huge amount of freely available intellectual property—reference designs, ready-ported operating systems, etc.—already extant for this core.

The cliché often used is that "ARM is the 32-bit 8051," meaning that it is the universal 32-bit microcontroller core known to everybody and used everywhere. This is barely an exaggeration; ARM is to the embedded world what x86 is to the desktop PC world.

It's important to keep your priority—low *overall* development cost—in sight at all times during the selection process. For example, I almost always reject parts that are only available in BGA packages, because it is practically impossible to hand-build prototypes around these devices, and it's costly to hire an external contract assembly house to build your initial development boards. You'll also need to consider the price and availability of evaluation hardware for the devices you're comparing, as well as the complexity of building a working hardware platform of your own. For example, a chip that requires complex analog support circuitry and careful PCB layout will be very difficult to work with in a hand-prototype environment. For such a chip, you would quite likely be better off investing in an expensive known-good evaluation board before attempting to build your own PCB. Diving straight into the deep end by designing your own board around such a part is likely to be costly, because of the need for several respins of your board to resolve layout-related and other analog issues.

Building Custom Peripherals with FPGAs

While you are evaluating different chips for your application, you are likely to find yourself tempted by specialized system-on-chip devices offered by various manufacturers. These chips will have interesting peripherals specific to various applications—for example, dedicated motion compensation and colorspace conversion hardware for digital video playback, or discrete cosine transform (DCT) engines for image compression, typical in devices intended for the digital camera market. Unfortunately, these are usually precisely the sorts of devices that are unobtainable to the hobbyist or small-scale developer. They are usually only available with solid up-front quantity commitments, and often non-disclosure agreements are also required. In some cases, just to view the datasheet for a part you will need to pay large fees to join some kind of specialized industry cartel (DVD/DVB playback hardware can be like this, for instance, because of the numerous patents in the field and vested copyright interests at stake).

Because of this annoying fact, one of the most useful money-saving skills you can acquire is experience working with synthesizable hardware design language (HDL) code on CPLDs and FPGAs. Using such devices, you can design your own custom peripherals, optimized for your specific application, and avoid the trouble of trying to source a rare ASSP. FPGAs are available off-the-shelf in many different packages and complexities, and in many cases the manufacturers supply free development tools.

In fact, there are now products available, such as Altera's Nios® and Excalibur™ devices, which consist of a high-performance RISC core "wrapped" in an FPGA, all on the one chip. Nios is a proprietary microcontroller core; Excalibur is built around a high-performance ARM922T core. With a part like this, you can effectively create your own custom ASIC; it is an extremely powerful tool and it seems likely that we can expect to see many more such devices in the future. ARM and other vendors also supply some cores in soft form, so you could in theory build your own entirely customized system-on-chip using a generic FPGA device. However, because of the hefty licensing fees involved, the per-unit break-even point is only reachable with very large production volumes.

If you plan to use FPGAs, much as with microcontrollers you will find that the manufacturer-recommended evaluation boards and commercial development tools can be very expensive. In the resources list at the end of this book, I make mention of Trenz electronic (www.trenz-electronic.de); this company is one possible source of lower-cost FPGA boards. However, you might not even need an evaluation board—FPGAs are, after all, *field*-programmable, and the interior functionality is controlled by the firmware you upload to them, so you can be fairly confident about dropping an FPGA directly onto a first-run prototype PCB and debugging your design in-circuit. If you've never used FPGAs before, however, I would advise getting a small evaluation board with which to experiment. Connect the I/O lines to pushbuttons, LEDs, or perhaps an RS232C level-matching IC like the Maxim MAX232A and play with the device to see what you can achieve with it.

Since I'm talking about field-programmable logic, I should also mention Opencores (www.opencores.org), an invaluable resource of free, open-source intellectual property ready to be compiled into your FPGA. If you need a core of some sort—a UART, for example, or a DRAM controller—then before starting to write your own, you should visit Opencores to see if there is already a free core available for you to adapt. Opencores is something like the Linux of hardware; at the time of writing, there are free cores for SDRAM controllers, UARTs, cryptographic hardware, microcontrollers, a VGA/LCD controller and many others.

Whose Development Hardware to Use—Chicken or Egg?

The textbook development cycle recommended by chip vendors is as follows:

1. Choose a microcontroller from the vendor's selection matrix.

2. Buy the vendor's evaluation board for this part.

3. Buy one of the commercial compilers, and possibly a hardware debugging module, recommended for the evaluation board.

4. License one of the operating systems recommended for the evaluation board.

5. Develop your application in vitro on the evaluation board.

6. Develop your hardware.

7. Port the operating system and your known-good application to the real hardware.

One of the driving ideas behind this methodology is that the software team doesn't have to wait for the hardware team to finish designing and debugging the circuit. Unfortunately, as with most textbook descriptions, the cycle described above ignores some important realities; not the least of which is that in many small shops, the job of both the software and hardware "teams" will be performed by a single person.

The evaluation board and software tools recommended by the chip manufacturer are usually expensive, for reasons touched upon in the introduction to this book. Additionally, if you intend to use complex off-chip functionality, it can be extremely difficult to attach this to an evaluation board. For instance, if you intend to implement a PCMCIA socket in your appliance, and the microcontroller evaluation board doesn't include one as an option, it could be hard to hand-build a PCMCIA interface board, and harder still to graft it onto the evaluation board. The majority of 32-bit parts are quite closely targeted at specific applications; evaluation boards tend to have all the hardware required to demonstrate the maximum possible bells-and-whistles configuration of the CPU's intended application, and this can get in the way of adding your own peripherals to the evaluation board. For example, I was once evaluating a chip targeted at the PDA market. The appliance I intended to build wasn't a PDA, so I didn't need most of the hardware on the evaluation board—audio I/O, Ethernet, color LCD, touchscreen, USB interface etc. Not only did I have to pay for all these peripherals (this particular evaluation board is US$1500, and the microcontroller itself only costs about US$12), but I had to cut several dozen traces, remove a 160-pin surface-mounted chip, and add literally a couple of hundred patch wires in order to be able to bolt on my own peripherals.

Finally, and following on rather neatly from the anecdote above, you should remember that the time required to understand the memory map and any special quirks of the evaluation board, and to get its specific combination of hardware running, is time that you are "stealing" from the task of getting your own circuit debugged. This is an acceptable price when you have a large team working simultaneously on the hardware and firmware of the final product, but in a smaller or even one-person environment working on a tight time budget, it is often more efficient to design your own circuit and start working directly on your own hardware.

There are three major ways around these problems, in roughly increasing order of difficulty:

- Locate a third-party demonstration platform for the part of interest.

- Locate a consumer appliance based on the chip that interests you, and reverse-engineer it enough to load your own firmware and patch on your own hardware.

- Design your own PCB and have it etched and populated either locally or (if this is a commercial project) by your factory; develop your firmware on this board while debugging the hardware at the same time.

The first option is rarely available, but usually well-supported by the board manufacturer. I should point out that in some cases it can be difficult to use these development boards unless you also possess a hardware debugging module such as a JTAG pod. Most difficulties center around how to upload initial bootstrap code to the board. Some microcontrollers, such as the Cirrus Logic CL-EP7212 and 7312 parts, contain a tiny on-chip bootstrap ROM that allows you to upload code to RAM over a serial port. You can implement your own flash-loader quite easily using this method, and thereby load your own code onto any board that has a serial port. Some evaluation board vendors will supply the board preloaded with a ROM monitor such as Angel or gdb stubs, and you can communicate with this monitor over a serial link. In a few instances, the board will feature socketed EPROM or flash memory devices, which you can simply remove and reprogram with your own code. Unfortunately, in a handful of cases, the board is shipped with blank, soldered-down flash memory and there is no way of getting new code into it short of buying a JTAG pod or some other specialized hardware device. Third-party "demo platforms" tend to be devices that were originally designed for some specific purpose, then later sold to hobbyists with no housing, but more detailed technical documentation. Easy field-reprogrammability with minimal external equipment may not have been a design criterion of the original appliance.

Repurposing consumer appliances can vary in complexity from extremely simple to downright impossible, depending on the microcontroller you're interested in and its target market. It can be exceedingly difficult to locate a consumer appliance based on the specific chipset it contains, and you will often need to do quite a lot of reverse-engineering in order to determine memory maps and so forth. It also isn't necessarily cheap to cannibalize a brand-new appliance, though it's almost always cheaper than buying an expensive evaluation board. The repurposing approach does have advantages for projects that meet certain prerequisites; in particular, it works best when you have a fairly good idea of the hardware capabilities you need (at a macroscopic level, e.g. "Must have Ethernet," "Must have TV output"), but you don't much care what specific parts are used in your hardware platform. As a result, this method is particularly attractive for hobbyist and student projects that are very price-sensitive and don't need to worry about ongoing component availability. People in this category can revel in the rich variety of items available on today's surplus market.

Between the years 1998–2001 in particular, seemingly dozens of companies—some big names, some unknown startups—developed many different styles of proprietary set-top box for various applications including interactive cable TV and living room Internet access. Over the same period of time, we have seen a proliferation of digital broadcast satellite service, digital cable TV, consumer DVD players and other digital media devices. These sources—particularly bankrupt vendors of proprietary set-top boxes—provide the secondary market with a rich supply of interesting and powerful hardware. Periodically, batches of these appliances appear on liquidation websites or in the "interesting surplus items" section of mail-order and online catalogs. At the very least, these items are often useful learning tools; in some cases, they can form the basis of a saleable product, an impressive student project or just a fun hacked appliance to have around your home. Although these devices usually contain at least some proprietary hardware (and/or code-protected microcontrollers that you can't read out or reprogram), they are almost always based around a well-known core, so if you can replace the firmware you can generally gain control of at least part of the appliance.

For interest's sake, I will describe a couple of examples that happened to be sitting on my workbench as I was writing this book. (Note: Don't expect to be able to go out and buy either of these specific appliances; I mention them, not as product recommendations, but strictly as illustrations of the type of hardware that frequently becomes available to hobbyists.)

The first example is the Newcom Webpal, an Internet-on-your-TV set-top box that, due to Newcom's dissolution, has recently been appearing on the surplus market in large quantities for around US$5 each. In fact, so many of these appliances have spread around in the hacker community that there is a significant amount of developer support for the product; for instance, a ready-to-run Linux distribution is available for download. While this device is perfectly usable as a general-purpose Internet appliance out of the box, it is more interesting for the re-useable hardware it contains:

- Cirrus Logic CL-PS7500FE microcontroller. This is a very powerful and flexible ARM7 system-on-chip device, originally designed for the Oracle® Network Computer platform.

- 1Mbyte of flash memory on a proprietary SIMM, with space for a second 1Mbyte chip.

- 4Mbyte of DRAM on a standard 72-pin SIMM.

- Infra-red receiver, remote control and wireless keyboard.

- Smartcard reader for ISO7816-2 form factor cards.

- Analog VGA, S-video and composite video outputs, and stereo audio outputs.

- Two serial ports (unpopulated).

- ISA bus with a small two-slot backplane, one slot of which is occupied by the CPU board. (The original retail Webpal had a 33.6kbps modem in the remaining slot; hackers have mostly replaced these with Ethernet cards).

Another recent example is the Virgin Webplayer. This is a small Internet appliance, very much like a laptop computer, though without a battery. It is essentially an attractively styled miniature PC with the following features:

- 10.4" 800x600-pixel color DSTN LCD.

- 233MHz National Semiconductor Geode microprocessor (x86-compatible) with CS5530 companion IC.

- 64Mb SDRAM on a standard 144-pin SODIMM.

- MiniPCI slot containing a 56Kbps modem.

- IDE and floppy controllers.

- Infra-red keyboard with integrated trackball.

- DiskonChip socket.

- Two USB ports.

This device was originally given out free of charge as part of the "Virginconnect" free Internet service; basically, Virgin expected to recoup their costs and turn a profit by tracking your Internet browsing habits and reserving parts of your screen for paid advertising. When the service was terminated, users were asked to return their units, but many didn't do so—large numbers immediately appeared for auction on eBay. Not long after this, the distributor of these appliances dumped vast numbers of the units on another online auction site. (In an amusing touch, the distributor's technical support staff also started directing customers with questions to a webpage that I had published several months earlier, describing the Webplayer's hardware, with driver downloads and other usage information.) These appliances are still in circulation at prices in the $125–$175 range, and judging from the email I receive, a large number have ended up as the heart of a student electronics project.

If you intend to work extensively with repurposed consumer equipment, I strongly recommend investing in a cheap JTAG pod such as the Macraigor Wiggler. Many microcontrollers have JTAG/ICE ports and appliance manufacturers using such parts almost always bring the relevant signal lines out to a header, or at least a set of pads for "bed of nails" test fixtures. This is done to facilitate post-assembly flash programming, factory tests and so on, but it makes your life a lot easier too.

Another tool I heartily recommend for this sort of work is Ida Pro (available from DataRescue, www.datarescue.com). This is an extremely powerful Windows-based interactive disassembler capable of inspecting and reverse-engineering code from a wide variety of microprocessors including ARM, Intel x86 and i860 (an older RISC platform that Intel is phasing out in favor of StrongARM-based parts), MIPS, SPARC, Motorola MC680x0 and Hitachi SuperH, as well as a few DSPs, and many 8-bit microcontrollers. If you need to reverse-engineer some firmware, this tool will make the job much faster and the final result more reliable. You simply load a binary ROM dump into the program, tell it which areas are code and which areas are data, and you can fairly easily generate a high-quality source listing of a device's firmware. You can scroll around inside the loaded program, following the execution flow or searching for particular strings or other data. For those of you who are familiar with the PC program Sourcer or the old Commodore-Amiga program ReSource, Ida Pro is conceptually very similar (in particular, it has almost exactly the same sort of user interface as ReSource); it just covers a lot more hardware platforms.

A note to non-US residents: If you live outside of the United States, you will probably find that the surplus channel is not quite as exciting as I've made out above. Unfortunately, many of the failed dotcom-style schemes that have led to really interesting hardware being liquidated at bargain basement prices are US-centric programs, and it's not usually possible to obtain the hardware elsewhere; surplus merchants are reluctant to ship overseas because of problems with credit card fraud, extra Customs paperwork required for such shipments, licensing restrictions, regulatory approvals and so on. However, even overseas it is well worth looking at local electronics surplus stores and catalog merchants. It's not uncommon for these dealers to acquire small quantities of appliances and/or replacement components from retailers, service centers and similar organizations who are going out of business or simply ceasing to stock or support a particular make of appliance. These bits and pieces are frequently sold under the catch-all of "unknown appliance—sold for parts." Just make sure that you order several at once of anything that appears interesting; if you buy only one, intending to take it home and reverse-engineer it, you'll almost certainly find that by the time you've decided whether or not the device is really useful, there are no more available!

The third development option, prototyping directly on your own circuit and debugging the hardware and firmware simultaneously, is the option I personally use most often. Although this method is common for low-speed 8-bit circuits, it is fairly rare in the development of 32-bit systems. However, I find it necessary to work this way because most of the projects I work on involve bringing together several fairly complex devices that aren't found together on any pre-existing evaluation platform. This method does have the advantage that you can tweak the hardware design to simplify firmware development right up until the last PCB revision before manufacture. Unfortunately, it also has the disadvantage that any bottleneck in the hardware development timeline is also a bottleneck in the software development timeline, which unavoidably pushes your delivery date further out.

I should warn you that prototyping like this is similar to bungee jumping; just one catastrophic failure, and you won't get a second chance. If you make a really fatal, unpatchable error in your PCB, in the worst case scenario you will have to throw it away (and more than likely the parts on it too; hand-reworked surface-mount devices have high failure rates) and halt firmware development until the next batch of boards arrives. This can make the process expensive, but with careful fault analysis and rigorous checking of your work before submitting a PCB layout for manufacture ("measure twice, cut once"!), you can keep the expense to a minimum. In a later section ("Special PCB Layout Rules For The Shoestring Prototype"), I discuss several rules you can follow to ease the process of developing this way and minimize your costs.

To summarize the above choices succinctly, then:

- **If your code can be developed on a readily available, affordable development board** (either third-party or direct from the chip manufacturer), you should use this development board as your prototype hardware platform.

- **If you are building a one-off piece** (e.g., a student project or technology demonstration), if you are *certain* you will never need to build more such units, and if you don't need to build around any specific component, your easiest route may be to repurpose a piece of consumer equipment with appropriate hardware features.

- **If you are designing around a specific component or combination of components, and the available evaluation boards are either too expensive or it isn't feasible to add the peripherals you need to them**, your best option is to design your own circuit, make a couple of prototype PCBs, and debug the application directly on your own hardware.

If none of the above options seems to be right for your application, then I suggest that you develop and demonstrate your software on an embedded PC type platform, and use this demonstration to secure sufficient funding to pursue one of the options above.

Our Hardware Choice—The Atmel EB40

For our fictitious project, I'm going to use the Atmel AT91EB40 evaluation board. This board is based around the AT91R40807 microcontroller, a simple 40MHz[3] ARM7 device with 136Kbyte on-chip SRAM and a modest collection of on-chip peripherals. The EB40 also features:

- 128Kbyte of flash memory (64K of this is reserved for a bootloader and the ARM Angel ROM monitor/debugger; you can erase the whole chip if you wish but if you do so, you will need external JTAG/ICE hardware to reload it).

- Two serial ports with RS232C-compatible level matching.

- 512Kbyte of SRAM (in addition to the AT91R40807's internal memory). This can be expanded to 2Mbyte.

---

[3] By default, the EB40 is configured to run the microcontroller at 32.768MHz. This text will assume that you have left the board at its default settings.

The EB40 is a superb tool for learning about the ARM series because it is based on one of the most popular ARM core variants (ARM7TDMI), and it is both very inexpensive and can readily be ordered on-line without needing to establish a project relationship with a distributor. It is also conveniently a "minimalist" evaluation platform, with a small number of on-board peripherals and a simple expansion interface, so not only is it inexpensive, but it's easy to add your own peripherals. In particular, if you combine the EB40 with an FPGA, you have a very flexible prototyping platform that can easily be turned into a manufacturable device.

Atmel also went to great pains to make it easy for almost anyone to get code onto the board; besides having the Angel ROM monitor included in flash (which can talk to the professional-grade compilers available from Green Hills et al as well as the free GNU gdb debugger), it has a JTAG/ICE port, and as an alternative route, the on-board bootloader can even load code directly into RAM over a serial port using a free Windows-based utility (BINCOM) from Atmel.

Recommended Laboratory Equipment

One question that arises frequently at this point is "What other equipment do I need to buy to equip my laboratory?". There seems to be a fairly widespread belief that developing high-end embedded systems requires a great deal of expensive specialized hardware; storage oscilloscopes, logic analyzers, in-circuit emulators and so on. While this equipment can sometimes be useful, the truth is that expensive state-of-the-art equipment is only absolutely necessary for a few special applications. For example, when developing cellular phones, in order to test your device without causing annoyance to local cellular carriers and the public, you need to be able to emulate a cellular network. In order to debug circuits that have extremely high-speed buses, or delicate RF or analog sections, you might also need some extra equipment, but for a large number of embedded designs, your needs are unlikely to exceed the following major appliances:

- **A reasonably feature-rich multimeter.**

- **A good analog oscilloscope.** Steer clear of generic no-brand entry-level scopes intended for the hobbyist market (even if you *are* a hobbyist). You'll find much better value in a refurbished piece of name-brand equipment. A quick search of the Internet will show you a large number of dealers who specialize in sales and rental of refurbished test equipment[4]. Brand-name units (Tektronix and Hewlett-Packard are the two most popular) that were state-of-the-art three to five years ago are now very affordable and more than adequate for most tasks. Your exact needs will obviously depend on what you're developing, but I would recommend a minimum 150MHz bandwidth 2-channel scope and 10x probes. Look for scopes with many triggering options—these options give you different ways of focusing on the specific section of the waveform you're interested in, and the more flexibility you have there, the better.

- **A laboratory power supply.** It should have at least two independently adjustable DC current-limited outputs (30V is the maximum you're likely to need), with inbuilt current and voltage indicators.

- **A bench-mounted illuminated magnifier.** This item is mandatory when working with surface-mounted parts, and it's useful even when working on larger packages.

- **A temperature-controlled soldering iron.** Always keep a few spare tips on hand, also—especially if you work with surface-mount packages, you will want to keep at least one tip filed to a very fine point. This point will erode quickly and you'll need to keep filing it down as necessary.

If you're working on something that will be powered from household wall current, and that you intend to distribute to other people, it's also a wise idea to have a variac on hand so that you can test how your device will behave in mains brownout conditions, but this isn't essential.

---

[4] You can also buy secondhand equipment from auction sites like eBay, but secondhand test equipment from a private seller frequently needs recalibration, especially after being shipped a long distance. It is often worth the additional cost to buy a certified, properly-packed unit from a reputable vendor of refurbished equipment.

Note that I haven't mentioned a digital oscilloscope. If you do want to buy one, by all means do so, but I suggest you make it a secondary purchase after acquiring a good analog unit. The main reason for this is simply cost; the same money will buy a much more capable analog than digital scope. Digital oscilloscopes are a time-saving luxury rather than an essential for many applications. I have a reasonably powerful digital scope on my workbench, and I rarely power it up. In fact, I most commonly use it when I run out of channels on my analog scope and I need to look at a large number of signals simultaneously.

I also recommend, in general, against the false economy of oscilloscope add-ons for PCs. The quality of the analog-digital converter side of these software/hardware packages is critical to the usefulness of the device. Expensive, high-speed data acquisition cards are outside the cost range of interest to the average reader of this book; cheap 8-bit digitizer devices with no internal buffering (typical of low-end PC oscilloscopes, especially those sold in kit form) are not money well spent, in my view. This type of hardware might be useful if you know you will be spending a lot of time looking at and storing signals at audio frequencies (up to a few tens of kilohertz); you can use the device as a poor man's logic analyzer. As a primary signal inspection tool, I feel this hardware lacks flexibility and, at worst, may be very misleading and counter-productive because it hides information that might be vital for debugging purposes.

Free Development Toolchains

A large majority of 8-bit and smaller embedded systems in the real world use proprietary (if any) operating systems[5], often written using a monolithic assembler/linker package. A great deal of literature for the embedded field deals with specifics about close-tolerance timing (cycle optimization of code) and single-byte memory-saving techniques. Professional debugging toolchains for these parts often center around using a hardware in-circuit emulator for the microcontroller to simulate the pro-

---

[5] "Proprietary" in this context means "developed specifically for one product or family of products," rather than the more general English meaning of "exclusively owned."

cessor in vivo, capturing and analyzing its behavior in realtime by means of an attached PC.

Design processes and priorities are usually very different when targeting 32-bit parts. To begin with, these parts are so fast that hardware emulators are unfeasibly expensive and almost all debugging is performed on the real microcontroller. (Sometimes, the microcontroller itself is used as a kind of in-circuit emulator using the JTAG interface. However, this serial interface is too slow for full realtime debugging.)

Also, particularly in the case of a demonstration or hobbyist project, the designer would probably like to avoid handcrafting all the code necessary to bring up a complex system, which implies that some kind of ready-made operating system will be used where possible. RAM and ROM are usually plentiful, making it unnecessary for users to spend a great deal of time squeezing a few extra bytes' efficiency out of their code. Algorithms are also much more complicated and have more points of interaction with each other and the external environment, requiring a significantly different style of design rigor.

As for cycle-exact performance issues, pipeline and cache features on these more advanced processors make hand-optimizing assembly language programs *extremely* difficult; in fact, instruction timing on a cached, pipelined CPU core under varying system load can be so complex that these systems sometimes actually appear to be nondeterministic. Optimization for speed is generally best left to a high-level language compiler on 32-bit platforms. Only if observed performance is inadequate and **actual profiler results** point to a specific area of the code is it generally worth the effort of hand-optimizing in assembly language.

Given these differences, which tools do we choose for our exciting new 32-bit project? With a few rather rare exceptions, the choice of embedded operating system will mandate the choice of a particular toolchain. Despite the proliferation of fairly well-defined binary file standards such as ELF, COFF and PE, differences in such compiler- and linker-specific behavior as symbolic debugging information, special directives for memory allocation, and C++ name mangling semantics usually make it very difficult to move operating systems from their intended compiler to an alien compiler. This problem is even worse with operating systems that are shipped partly or wholly precompiled, without sourcecode. Although it is possible, in some cases, to force specific combinations of products to work together (e.g., object files compiled with the ARM Developer Suite can be massaged to link with code generated by gcc), this is rarely a wise expenditure of time.

Keep in mind that this interrelationship works in reverse too—in other words, if you don't want to spend the money on a costly commercial toolchain, this is probably going to limit your choice of operating systems. For the purposes of this book, I am going to consider all the commercial tools as being too expensive (licenses for this type of product typically start at around US$3,000), so we are going to focus on platforms that are supported by free compilers. For all practical purposes, this means platforms supported by the GNU tools; gcc et al. There exist a few free, manufacturer-supplied proprietary compilers, but these vary widely in quality and are generally nonstandardized. Unless your chip or operating system vendor is going to supply you with a huge variety of free, useful intellectual property in the form of libraries that can only be linked with the proprietary compiler and for which you can't obtain open-source equivalents, I strongly advise that you stay on the far better-traveled path of GNU tools. It's hard to imagine any algorithm from cryptographic applications to video decoding for which GNU or other open-source intellectual property isn't already available. Freely available source probably won't be optimized for your hardware platform and will require some tweaking for best results, but even so the benefit of having the sourcecode is very significant.

I should pause here to point out that if you are using the Intel x86 family for your platform, there are at least two other viable free compiler options for you. Borland has released the command-line version of Borland C++ 5.5 as a free download, and the Watcom C++ compiler (now owned by Sybase®) is in the process of being released as an open-source product named OpenWatcom (www.openwatcom.org). OpenWatcom is not available for general download at the time of writing, but when it does finally make it to the outside world, it should be a very exciting product. Watcom C++ supports numerous Intel targets—

Win32, Win16, OS/2®, Novell® NLMs, and both 16-bit and 32-bit DOS. With a little external massaging, it can be used to develop almost any x86 code for embedded platforms, especially when combined with a free operating system like FreeDOS (www.freedos.org). In the heyday of DOS, Watcom C++ was also famous for generating highly speed-optimized object code for DOS-based games, which may be an interesting advantage for your application.

The GNU suite is a software-only toolchain, meaning that we need to establish our own link to the target hardware for code uploading and debugging. Most of the 32-bit parts we mention in this book, and certainly virtually all ARM-cored parts (including ASICs), include on-chip JTAG hardware debugging support. For those who haven't used it, this is a simple serial interface that allows external hardware to halt the processor core and inspect and manipulate its state[6]. Through this mechanism, it is possible to generate read/write cycles that appear to originate from the core, and thereby operate and examine on-chip peripherals and external hardware. In order to make use of this interface, you need a JTAG pod; these range in complexity from fully autonomous standalone units that connect to your computer via Ethernet to simple devices that level-match and buffer your PC's parallel port signals onto the target's JTAG port pins. The only readily

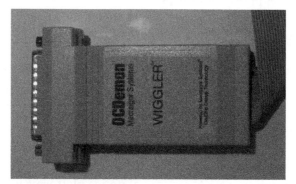

Figure 2-1. Macraigor JTAG Wiggler

---

[6] The JTAG interface can, of course, be used to manipulate other on-chip hardware directly. However, doing so would require device-specific knowledge of the on-chip peripherals. By taking control of the core, we can generate read and write cycles to access other system hardware without proprietary knowledge of each different microcontroller.

available JTAG pod I have found that lies within a shoestring budget is the Macraigor Wiggler, illustrated in Figure 2-1.

The Wiggler belongs to the category of simple parallel port devices; however, it is an extremely powerful tool. With it, you can halt the processor and inspect its state, as well as being able to read and write hardware registers and other memory locations. This capability will save you a lot of time when you're working out how to bring up a new system; instead of having to recompile, upload and test your bootstrap code iteratively, you can simply connect the debugger and tinker with the hardware registers directly until the peripherals are behaving the way you want them to. Moreover, because the JTAG interface is entirely hardware-based (on the microcontroller end), you can use it to breathe the "kiss of life" into a board with blank flash memory.

There are a few hardware projects that duplicate the Wiggler's functionality (it's a very simple device), but the really tricky part is not the hardware, but learning the scan chain codes for the chips you intend to debug. This information is usually closely guarded by the chip manufacturers, and you really need to be a large corporate entity in order to have access to it. For this reason, I recommend sticking with a hardware vendor like Macraigor who has good relationships with the chip vendors, to ensure ongoing support for new parts.

Free Embedded Operating Systems

Having introduced the common choices for free development tools above, let us explore some of the operating system choices available to us. Fortunately, the open source movement has generated a plethora of free, or nearly free operating systems, probably the best-known of which is Linux. One great advantage of Linux is that not only has it been ported to a great many architectures, but the install process for many reference platforms is relatively well-documented. Being able to download a working, precompiled kernel and fairly precise installation instructions will save you an enormous amount of frustration at the start of a new project. In the last year or two, Linux has also attracted quite a lot of attention from the embedded world, and as a result we are starting to see some embedded-specific features emerging in the mainstream

Linux code. For example, current kernel versions directly support ROM-based filesystems (including compressed filesystems) as well as several forms of flash technology including NAND flash (SmartMedia et al), M-Systems DiskOnChip, et al.

For some applications, it may be valuable to note that "pure" Linux has three important limitations:

1. It requires a hardware memory-management unit (MMU) in the target processor.

2. It is not, strictly speaking, a realtime operating system.

3. It is licensed under the GNU General Public License (discussed in the next section of this book), which may have privacy implications for your own code.

The reason I qualified the second point above is because off-the-shelf Linux can often be thought of as "realtime, for small values of realtime". In other words, stock Linux may be quite realtime enough for your needs, especially if you are willing to massage the kernel a little. Developers who are accustomed to working with actual real-time operating systems will doubtless cringe at my cavalier treatment of this issue, but for many non-critical applications, simply using a fast enough processor and removing unnecessary background tasks will be sufficient to ensure that your application gets enough processor time to *appear* to be working in realtime. The difference between this and a true RTOS is that the RTOS will have APIs to *guarantee* that, for example, a level 0 interrupt will be serviced within 2ms of the hardware receiving the interrupt request, or that a given process will always get at least 25ms out of every 100ms of processor time. Whether or not you can get away with a non-realtime operating system depends on your application; principally, if physical or financial safety depends on your appliance being truly realtime, then you *must* either use a true RTOS or modify your existing OS so that you can guarantee that any critical code will be allowed to run when it needs to.

If you need a truly realtime version of Linux, there are a few options open to you, but probably the best-known is a commercial distribution called Hard Hat Linux from Monta Vista Software (www.hardhatlinux.com). Monta Vista also makes a specialized version of Hard Hat Linux for the telecommunications industry. Another option, and one that you can download freely, is RTLinux (available at www.fsmlabs.com). Despite what you might be told, the "realtime" versions of Linux are not really fundamentally different operating systems; they essentially consist of a small realtime subsystem melded to a normal Linux system. If your realtime needs are modest, you may be able simply to add your own minor patches to the Linux kernel to run your own critical tasks when necessary, rather than inheriting any idiosyncrasies of someone else's "two-pronged-kernel" realtime Linux design.

If you need a version of Linux that will run on microcontrollers lacking a memory-management unit, there is also a version to accommodate you: ucLinux (www.uclinux.org). ucLinux is a public project with a strong leaning towards projects that involve repurposing existing appliances such as Palm PDAs. The ucLinux website also features links to some interesting, moderately priced hobbyist 32-bit development boards based on processors such as the Motorola Dragonball series.

Without a doubt, Linux is the operating system de rigeur in the hobbyist arena. Partly because of the percolation of hobbyists into the commercial world, and partly simply because of the operating system's own merits, there is large and growing commercial use of and support for Linux-based embedded solutions. For some examples of this support, you should visit www.linuxdevices.com, which is probably the most comprehensive portal site for news of the embedded Linux world. There are a surprising number of product announcements from major vendors aiming at the consumer electronics market. Linux's position as the server operating system of choice on the Internet seems to have helped to make it the top contender to run the next generation of networked home entertainment and other appliances. (It's also well worth visiting linuxdevices.com when you're searching for a ready-built hardware platform for some embedded application or even just for prototyping purposes. The site contains numerous interesting articles and product lists for various embedded computing platforms that can run Linux, and of course there is no reason why you couldn't load your own operating system onto one of those boards. Some reviewers of this text have pointed out to me that there's almost nothing at this portal site that you can't find by some reasonably diligent Web searching, but after all the primary purpose of a portal is to collect audience-targeted information into one convenient location so you don't have to do the searching legwork yourself.)

Another popular free UNIX variant is NetBSD (www.netbsd.org). This operating system has one major advantage over Linux: it is unconditionally free[7]. Like Linux, NetBSD has been ported to a huge variety of platforms, and supports a wide range of miscellaneous hardware. The main disadvantage to NetBSD is that it has not attracted very much attention from hardware OEMs, at least compared with Linux. The Linux community is sufficiently large and vocal that hardware vendors generally provide at least token support, whereas NetBSD is a poor cousin, relatively speaking. There is a fair amount of code interchange (within licensing limits) between NetBSD and Linux, and a large number of Linux projects can be rebuilt on a NetBSD base, but overall if you are looking for sheer breadth of ready-made hardware drivers and availability of peer support, Linux is probably a better choice. However, if it is important to you to keep every line of code you write secret, then you should look more closely at NetBSD. Although, with due care and attention to licensing details, you can build a Linux system that doesn't require much (if any) disclosure, you may find it easier to get NetBSD past a reluctant management team who has been frightened by or is otherwise doubtful about the legal status of open-source projects. You can simply tell your managers that NetBSD is unambiguously free, there is no disclosure of sourcecode required, and that will (hopefully) be the end of those managerial objections.

Linux and NetBSD are both very "heavy" operating systems; they require a relatively large amount of RAM and nonvolatile storage space (either ROM, flash memory, or another device such as a hard disk). This can be mitigated to a certain degree by very carefully pruning the kernel and deleting unnecessary binaries and libraries, and by using special slimmed-down system librar-

---

[7] Although the NetBSD operating system kernel is covered by a virtually unrestricted free license, individual components of a distribution may be covered by different licenses such as GPL.

ies, but neither product was originally designed for embedded systems. Both products are also very flexible general-purpose operating systems, and of course this flexibility comes at a price.

A slightly lesser-known free operating system, but one with growing popularity, is eCos from Red Hat (sources.redhat.com/ecos). The great advantage of eCos is that it was purpose-built from the ground up as an embedded operating system, unlike Linux and NetBSD. Although it is monolithic in the sense that it compiles into a single library that you link with your own program (as opposed to being a heterogeneous collection of executables, configuration files and libraries that need to be stored in some kind of filesystem), eCos is a very well-designed modular operating system. The presence or absence of drivers for various hardware, and all configuration options, are controlled easily with conditional compile macros. RedHat even includes the unaccustomed luxury of a graphical configuration editor that lets you set all the build options with checkboxes, drop-down lists and so forth, and build the operating system library with a single keystroke.

eCos can also be compiled for operation from RAM (extremely useful for debugging; you leave the ROM monitor in control of the board and simply upload new versions of your application as you debug it), ROM (useful when you go to burn the firmware into your device!) or a combination of RAM/ROM startup, where the code is initially located in ROM, but relocates itself to RAM for performance reasons. The operating system is supported by a highly flexible bootloader called RedBoot; this bootloader is a very interesting product in its own right, since it offers a simple command-line loader accessible over serial or Ethernet (where supported), flash rewriting commands, and other useful functionality.

At the time of writing, there are basically two publicly available versions of eCos and its support tools—an ancient "official release," and the current CVS version. (CVS is a version-control tool commonly used in the free software world.) If you intend to play with eCos, download the current CVS version. Instructions for doing this can be found at the eCos web site, sources.redhat.com/ecos. The "official release" version is ancient;

the CVS version, though it is something of a moving target (since it is not a frozen version, it changes frequently) supports many more hardware platforms and has many more features than the old release. If you're using Windows, however, I do suggest you download and install the official eCos release and then update it with the latest CVS version. By doing this, all the necessary default configuration information, registry values and so on can be initialized by the automated installer. Be sure to read all the download pages carefully, however—all of the old utilities supplied with the release version of eCos must be updated manually with newer versions if you are using the CVS version of the operating system sourcecode.

Another operating system which isn't truly free, but is *effectively* free, is the Palm OS. The reason I describe it as "effectively" free is that the only way you're likely to be using this operating system in a shoestring-budget project is if you're implementing your project as an application running on a dedicated Palm device (or third-party compatible; Sony Clié, Visor, IBM WorkPad, etc). Since the operating system comes bundled with the hardware platform, and free development tools and documentation are available, shipping applications based on this OS is basically free. In fact, quite a few niche market products work precisely this way; you pay for an off-the-shelf Palm device preloaded with custom application software, and possibly some special external hardware such as a GPS receiver, barcode reader or digital camera. An obvious advantage of implementing your project in this way is that as new and more powerful hardware platforms become available, you can upgrade to them quite painlessly; Palm will handle all the work of porting their operating system to the new hardware and you can reap the benefits. (A similar situation applies to Windows CE. At least at the time of writing, you can download free Windows CE compilers at Microsoft's web site). This technique, however, barely falls under the heading of embedded systems development, and so I will not discuss it further in this text.

Of course, depending on what functionality you require, it might not be necessary to port and bring up an entire operating system just in order to acquire some ready-rolled functionality.

Some vendors provide modular packages for specific functions (these are usually supplied as precompiled libraries, so make sure that they can be linked with your toolchain of choice). For example, US Software (now owned by Lantronix) sells standalone modules for TCP/IP networking (USNet®) and DOS/Windows-compatible VFAT filesystems (USFiles®), in addition to several embedded operating systems.

It might also be feasible for you to "mine" small fragments of code out of an existing operating system and create your own libraries. If you are thinking this latter is the best route for your own project, remember that as the size of the code piece you're extracting increases, so do the number of structural assumptions you're inheriting. For example, if you want to borrow a filesystem driver out of an operating system, you will either have to modify it heavily to fit your own code, or you will have to duplicate the file descriptor semantics at the top end, and the low level disk-access device driver semantics at the bottom end, not to mention task synchronization primitives and so on. Effectively, you may find yourself emulating or rewriting large segments of the operating system from which you borrowed your "single" piece of code.

Remember also that even if you start out building a prototype around a ready-made OS, it is entirely possible to "wean" your code off that OS at a future date—though it will be much easier if you start out by designing your code with this intention in mind (see the section headed "Reliability and Portability Considerations" later in this book for more information on this topic). When you start a new project from scratch, it is very helpful to have some piece of code around, even if only for reference purposes, that you can trust to work properly. This is especially valuable if that piece of code can teach you the correct method and order of initializing the components of a complex system. For example, I once worked on a project based around an ill-documented Super-VGA controller IC. The chip vendor actually supplied free reference sourcecode to bring up the SVGA chip, but it wasn't complete and didn't work. Fortunately, they also provided a working RTOS preloaded on the evaluation board. I obtained the necessary magic register values to get my own code working by booting up the vendor's proprietary RTOS, letting it initialize the display control registers, and then dumping the entire chip state (including, as it transpired, many undocumented registers!) to a serial port for inspection.

Because of the possibility of issues like this, you might want to use a ready-made operating system (on your real hardware) to get your application up and running quickly, and gradually replace parts of that operating system with code of your own until eventually you have duplicated all the desired external functionality in your own application. This is an exceptionally valuable method of doing things when the only operating system that explicitly supports your reference platform has expensive royalty fees, but is free for in-house research use. You can simultaneously cut your per-unit costs and your development time by starting your program out as an application on top of the expensive OS. Once you've determined what services you actually need out of the operating system, you can go through your code replacing all the operating system calls with your own hand-written versions of the same functionality. Once you're done, you have a shippable proprietary application that doesn't use any of the expensive third-party code. Obviously, this is a lot more work than simply writing your program around a ready-made operating system, but on the other hand it does save you a lot of debugging work in the initial bring-up stage, and it avoids potentially large operating system license fees.

Note that this technique is subtly different from the technique of prototyping your application code on some arbitrary hardware platform, with the intention of porting it to real hardware once the algorithms have been verified on the demonstration hardware. Using the method above, we are developing on our real hardware (or at least the reference platform we are using to develop the real hardware). At any point, we could bundle together the current codebase, load it onto a piece of real hardware and call it a shippable product (at least from a functional perspective); the only delay is caused by the need to remove expensive licensed code. By contrast, the in-vitro code prototyping system doesn't result in a shippable product until the very end of the prototyping and porting process.

In the simplest case, you might not need to use an operating system or third-party libraries at all; you can roll your own entirely proprietary code. The toolchain you will be building later in this book works well in this type of scenario; it has a reasonably powerful C run-time to save you from the drudgery of writing functions such as memcpy() and sprintf(), but it is entirely OS-agnostic.

GNU and You—How Using "Free" Software Affects Your Product

In the modern era, almost any nontrivial embedded project of the type we are discussing will require an enormous volume of essentially boilerplate code; TCP/IP networking, data compression, filesystems (particularly MS-DOS-compatible FAT filesystems - "Where can I get code to read a FAT-formatted hard disk?" is a frequently asked question in embedded newsgroups), audio/video codecs and GUI libraries are common examples. Of necessity, therefore, implementing such a project from the ground up involves reinventing many wheels. At the very least, this is an inefficient use of your expert time. At the worst, it can mean a project that never gets off the ground because you don't have the manpower needed to get the pedestrian code finished so you can move on to building the value-added magic that makes your product something special and saleable.

In the past, these unpleasant facts could be worked around only by purchasing expensive commercial RTOS packages. However, in recent years, many free alternatives have become available and viable, and the use of open-source[8] "free" software in commercial ventures has been greatly legitimized. Despite this, there is still a state of confusion in the minds of many embedded engineers and entrepreneurs alike as to just what it means to use open-source software; what rights and benefits it confers, and what obligations it entails. This situation is not ameliorated by the fact that most of the outspoken experts in this field are vigorously pursuing commercial or political agendas and in many cases intentionally obscuring the facts. In order to fully understand the implications of using some of this free software, it is therefore

---

[8] "Open source software" is a politically loaded term with multiple more or less widely accepted meanings. In this context, I am using the phrase to mean "royalty-free software for which the complete sourcecode is readily available without payment of fees."

necessary to be armed with at least small amount of background information about this political situation. Please note that this is intentionally only a brief description, and of course it constitutes neither formal legal advice nor a complete analysis of the social and legal issues surrounding any particular software license.

The reason you need to read this chapter is that when you're implementing a complex project, sooner or later you will be forced to choose between a proprietary operating system or a free product covered by some kind of "open source" license. Chances are good that you will be facing one or more salespersons and free software advocates, each of whom will not necessarily present you with complete information to make your decision. Depending on your organization's structure and history, you may also be combating misconceptions in management about the implications of using "free" software in your product. Free software, used properly, can be part of any totally reliable, legally sound, high-performance product; this approach to software development can no longer be considered trailblazing, and it remains only to select which type of free software you should be using.

The most popular free software license (in terms of lines of code freely available on the Internet, at any rate) is unquestionably the GNU General Public License, commonly abbreviated "GPL". Most Linux software, for instance, including the Linux kernel itself, is released under this license, and most free software controversy in the public press centers around GPL. The rationale behind GPL, simply stated, is to force all derivative works of open-source products to remain open-source. (The actual rationale goes somewhat deeper than this; it is based on the idea that all software should be free, in the philosophical sense of the word; a true free software purist abhors the concept of closed-source applications.) The two aspects of the GPL that will affect you most are:

1. You can experiment with GPLed software as much as you want in private. You only "accept" the license and therefore become bound by its provisions once you "distribute" products derived from GPL code.

2. If you distribute a product that is derived from or closely linked to GPL code, your code must also be released under

the GPL. This means that you must release sourcecode (or disclose a means of obtaining the sourcecode) to anyone who requests it. There is an important exception to this rule for the Linux kernel: You do not need to GPL a piece of software whose only link to the Linux kernel is that it calls kernel services using documented interfaces. The original intent of this rule was to allow people to develop Linux device drivers for products whose hardware documentation is covered by nondisclosure agreements (an intent largely nullified by later philosophical changes in the license), but it also provides a very useful way of allowing profit-making use of the large amount of engineering in the Linux kernel.

For in-house prototypes and private experimental research of all kinds, the first rule above is a largely unrestricted free ride. You can take an existing mostly-GPL project (like a Linux distribution) and use it as the foundation for your prototype without restrictions. Once you're satisfied that your code and/or hardware are working nicely, you can decide exactly how to bring the product as a whole to market and remain license-compliant. However, you should plan now for what you intend to do when you commercialize your product. Otherwise, you'll demonstrate a fantastic but legally unsaleable prototype at a trade show, people will come to you ready to write orders, and you'll have a huge auditing and rewriting job before you can cash their checks. Your options are as follows, in ascending order of person-hours typically required:

- Release your entire product under GPL. This option can make a lot of sense, particularly when your product is largely special hardware that just happens to require control software (as opposed to general-purpose hardware running special software, where all the value lies in the bundled software). If you take this route, you can also ride a certain amount of bonus publicity from the free software movement, who will be only too happy to promote your product as an example of embedded engineering done right. This extra goodwill can be very useful in some markets. However, sometimes there can be other issues—typically, nondisclosure agreements required by other product vendors you work with, patents and various other trade secret problems—that preclude this option, even if you are personally willing to try it.

- Establish a clear separation between GPL and non-GPL code in your product, and open-source only the GPL components of your software bundle. This technique is exceedingly useful when your product is based around Linux, because the Linux kernel exception to rule (b) mentioned above gives you a convenient place to draw the "GPL vs. non-GPL" line in your software bundle. The Linux-based TiVo digital video recorder appliance and Sharp's range of Linux-based PDAs (such as the Zaurus SL-5600) are excellent contemporary examples of this technique. All you are required to release are the special device drivers and other kernel modifications you may have written to get Linux up and running on your hardware; your application code remains secret.

- Determine exactly what GPLed functionality you're using, write your own implementation of all that functionality (or buy someone else's proprietary implementation), and remove all GPLed code from your software bundle prior to release. This is obviously the brute-force approach. I've listed this option last because it is usually the most labor-intensive, but this isn't necessarily true for all applications. If you're very careful to maintain an abstraction layer between your application and external libraries and operating system calls, or your application is of such a nature that it doesn't require many external services, this option might be the best for you. However, the applications that are easy to "de-GPL" in this way are precisely those applications that probably wouldn't have required importing a whole operating system in the first place.

Besides the special rules for the Linux kernel, there are some other varieties of GPL. One of the most useful is the "LGPL," which originally stood for Library GPL but is now referred to as the Lesser GPL. The LGPL is very similar to the Linux kernel license, except that it refers to a single library rather than the kernel itself. Libraries that are licensed under the LGPL can be used by your program without triggering a requirement to GPL your own code, as long as you only use documented calling mechanisms.

One of the most common licensing cases that people ask about in embedded discussion forums is exactly how they can build Linux into their system without having to release all their sourcecode. The answer to this is that the code you write will fall into three categories, with different licensing implications for each:

1. Kernel modifications. This includes patches you have made to the public sources as well as additional loadable kernel modules you may have written. Source code for these must be disclosed.

2. Modifications to LGPL libraries. You will need to disclose all your sourcecode for these.

3. Your own application. As long as you only use the kernel's documented interfaces, and documented interfaces to any LGPL libraries you use, your application code can remain secret. You must not make use of any libraries or other modules that carry a full GPL license, or you trigger a full GPL disclosure requirement on your own code. You must also avoid any undocumented interfaces to LGPL libraries or the kernel itself.

In practice, a large majority of embedded Linux projects will use exactly one library—glibc, or a cut-down variant of it such as uclibc—without ever needing to modify it, so the caveats in cases 2 and 3 above are never encountered.

At the opposite end of the spectrum from GPL, you will find the NetBSD license. This is a refreshingly simple license, which allows you to download the free sourcecode, experiment with it, use it and release derivative products, with or without sourcecode disclosure as you see fit. The only real limitation is that your product and its advertising materials must acknowledge the original author, typically with a phrase such as "This product includes software developed by X". (Some variants of the NetBSD license have dropped this last requirement.) There are also some common-sense requirements which are in no way onerous: you agree not to use the original author's name to promote your derivative product, and you agree that the code you received has no warranty. The NetBSD license is literally something for nothing; you get the sourcecode for free, you can distribute your binaries and charge money for them if you wish, and there is no requirement for sourcecode disclosure (though of course it is encouraged to release as much as you can). A splendid example of NetBSD in widespread commercial use is Apple's latest generation of Macintosh operating systems.

There are innumerable other open-source licenses, many of which are associated with just one specific product. For example, the Red Hat eCos operating system described in the previous chapter is released under the "RHEPL" license similar in philosophy to the NetBSD license. All of these miscellaneous licenses lie in a spectrum roughly bounded by GPL at one end and NetBSD at the other (in terms of sourcecode disclosure requirements vs. recognition of proprietary trade secret rights), with special conditions in some cases. However, you will find that the majority of the interesting open-source projects in the world are GPL-licensed.

One incidental pitfall does bear mentioning: There is a surprisingly large amount of open-source material which implements patented algorithms. For whatever reason—be it a love of academic freedom of speech, a desire to avoid expensive legal action, or simple lassitude—the owners of these patents often don't see fit to enforce them for free products. Even if you comply with the license agreement for the freeware product, that does *not* imply that you have somehow inherited a right to use the patented intellectual property in your project. For instance, there are freeware DVD playback programs readily available on the Internet. (For the benefit of those who know about such things, let's leave the thorny issue of DeCSS and the evil MPAA out of the equation and consider only an unambiguously legal freeware MPEG-2 player capable of playing unencrypted, legal DVD content such as you might produce if you use a consumer DVD recorder to convert your home movies from VHS to DVD format.) Notwithstanding the free nature of the code license, if you use one of these players as the core of your own consumer electronics DVD player project, you'll find the DVD consortium knocking on your door very quickly indeed looking for monies related to use of the DVD video trademark. You'll also be facing litigation on a raft of patent issues surrounding the MPEG-2 decoder. Several jurisdictions are currently evaluating possible changes to the way software patents are granted, with the abolition of such patents (or at least limiting them to a few years) being one of the options under consideration. Until such an enlightened, forward-thinking step occurs, however, you need to be willing to research possible patent protection of the the algorithms used in your project, regardless of what the licensing conditions might be on any particular implementation that you have referenced in your code.

Whatever third-party intellectual property you wind up using—even if you don't include third-party code in your final product release—it is absolutely essential to maintain an audit document for your software. This need not be a particularly onerous task; at its simplest, it can be a document listing each item you have included in your product (operating system kernel, third-party libraries, example sourcecode, clip art, fonts and so on) along with a copy of the license agreement that accompanied each of these items when you obtained them. This latter is particularly important because some licenses evolve over time (GPL is an example)—the license you obtain today may not be the same license that you would obtain by downloading the software tomorrow. With this document in hand, you have a documented legal defense against any accusations of license violations.

One last note: With the plethora of useful open-source code floating around the Internet, free for the downloading, there might be a temptation simply to download and use whatever you please and assume that nobody will ever know because nobody will ever see your sourcecode. Even ignoring the moral issues, this is suicidal folly. Anything from a disgruntled (or simply talkative) staff member to an interested hacker to a competitor reverse-engineering your product will destroy your company; discovery is inevitable, particularly if your project turns out to be a success. At the time of writing, several major American corporations are writhing in the throes of government investigations into accounting fraud; if your major product contains plagiarized code, discovery will lead to similar consequences. Worse—and this also applies to privately held companies, because it's not just a stock price issue—you may be unable to ship any more units without an expensive major rewrite of your operating system. Don't take this kind of risk. If you use free code, honor the license.

Choices of Development Operating System

Most hardware tools—PROM burners, programmable logic interfaces, debugger modules and so on—have proprietary hardware interfaces and are usually only supported under Windows and/or commercial UNIX variants such as Solaris. Many programmers, particularly those without much experience on high-end embedded systems, are also more experienced with working inside Windows. In addition, you may have other tools—CAD software, for example—which is only available for commercial operating systems like Windows, and to which you will be referring frequently.

Unfortunately, the open-source movement, or at least the GNU project (which provides us with high-quality free development tools) is focused more on supporting free UNIX variants such as Linux and NetBSD. Although of course you can have a dual-boot system, or multiple PCs on your desktop, this can be irksome—particularly the dual-boot system.

The simplest answer to this problem, at least for Windows users, is to use the Cygwin environment from Red Hat. A recent version of Cygwin is included for your convenience on the CD-ROM with this book. This piece of software allows you to run a simulated UNIX environment within Windows with minimal performance overhead. Detailed instructions for installing Cygwin are included later in this book. Cygwin is probably the easiest way to get up and running with GNU tools inside Windows; it is certainly the most popular. This description would, however, be incomplete if I failed to point out that Cygwin is a class of product sometimes referred to pejoratively as a "butterbox"[9]; in other words, it's something of a hack. Cygwin is probably the best solution if you must run Windows for other reasons, but it can be quirky. (However, if you are using the Macraigor Wiggler JTAG debugging pod mentioned earlier, remember that the software for this device is only supported under Windows. Linux support for the product has been rumored to exist because a third-party

---

[9] This phrase refers to a feature found in some refrigerators; a heated box inside the cooled compartment, designed to keep your butter soft and spreadable. Generically, it refers to software with seemingly useless or contradictory functions, e.g., a DOS emulator for DOS.

developer at one time obtained the control specification from Macraigor, and rolled it into the gdb sources. This code is not part of the publicly-available gdb source tree.)

Note in particular that Cygwin works best inside NT-class versions of Windows (NT, 2000 and XP). Due to differences in the executable loading behavior of DOS-based Windows variants (95, 98 and Me), build performance is significantly slower under these operating systems. Some extremely complicated projects, such as gcc, may not build correctly at all inside Windows 95/98/Me.

There is another project with similar goals to Cygwin, called MinGW (Minimalist GNU for Windows). The main technical difference between MinGW and Cygwin is that MinGW executables are standalone; they use Windows system services directly, rather than going through an abstraction/emulation layer like Cygwin. In other words, the translation from UNIX to Windows APIs occurs at compile-time with MinGW, rather than at runtime as with Cygwin. It's certainly possible to build cross-compiling versions of the GNU toolchain that operate with MinGW, but for the moment Cygwin is the mainstream route for people who prefer to work in Windows.

It's even possible to build GNU toolchains hosted on DOS (with the DJGPP 32-bit extender), but there are few good reasons for attempting this.

Yet another way of approaching the problem is to run an alien development operating system inside a complete hardware emulation of a PC, using software such as VMWare, or Connectix Virtual PC. For example, you might choose to run Windows XP as your primary desktop OS, but develop inside Linux running within such a software emulator. There are some minor advantages to this—for instance, it makes backups easy (since you only have to back up a virtual hard disk file), but there is a substantial performance impact and this approach is not recommended.

Overall, I recommend using Linux, if possible, as your host operating system when developing embedded projects with the GNU toolchain. By developing under Linux, you will simultaneously avoid the quirkiness of Cygwin, the performance issues of full virtual-machine emulators like VMWare, and any issues you might encounter using a less-popular system such as MinGW or DJGPP. As a side benefit, you will be using an entirely free development environment, which if nothing else is consistent with the goals of this book.

Special PCB Layout and Initial Bring-Up Rules for the Shoestring Prototype

Perhaps ten years ago, most parts of interest were available in through-hole package variants and it was usually possible to assemble most prototypes on simple off-the-shelf 100mil perforated PCB material, Veroboard® or wire-wrap boards. These materials are quite cheap, and it is easy to build such devices with only reasonable care and manual dexterity. For this reason, many introductory engineering courses and most hobbyist projects still deal exclusively with large through-hole parts. However, the high pin count of most 32-bit parts (and their support chips; SDRAM, high-density flash memories, display controllers, Ethernet controllers and so on) means that they are usually only offered in surface-mount packages such as TSSOP, QFP and BGA. This makes prototyping (especially by hand) much more difficult and rather more expensive, but it is still within the capabilities of a hobbyist workbench as long as you follow some fairly simple rules. Even if you are basing your development around a pre-built platform such as an evaluation board, you will almost certainly have to make your own PCBs to mount your peripheral components, and the rules below are equally applicable to peripheral boards as to complete systems.

The first and most important rule is that all programmable components containing nonvolatile memory (microcontrollers with internal flash memory, EPROMs, flash memory, CPLDs, etc.) should be socketed. The only exception to this rule is if you have a means of loading new firmware onto the device that will work while it is in-circuit *and the rest of the board is nonfunctional.* Although you can rely on the microcontroller to perform flash reloading operations once the board and at least part of the firmware are known good, that doesn't help you debug the hardware and develop the first version of your bootloader!

This may be a problem if your design uses large or wide flash memory devices. These parts are only available in TSOP/SSOP and similar surface-mount packages, and sockets for those packages are extremely expensive. The easiest route around this particular issue is to put a JTAG header on the board and solder in the flash memory, but if this option isn't available to you (if your processor doesn't support JTAG, or if you don't own the requisite interface module), probably the easiest thing to do is put a proprietary header on the board, wired in parallel with the footprint for the real surface-mount part. You can then build an external flash memory bank using through-hole parts and connect it to the proprietary header. For example, if your design calls for a 1Mx16 flash part, you can build an off-board module containing four socketed 512K x 8 parts (which are readily available in DIP packaging and can cheaply be socketed) and some address decoding logic. Although this will obviously not be compatible with write algorithms for the 1Mx16 part, you will be able to use it to develop your bootloader, and possibly most of your application.

The tricky part will come when you need to develop the code that writes back to that flash memory. It's hard to offer a truly universal generic suggestion for overcoming this problem, but the way I normally achieve this is by routing the chip select lines of the "special memory module" header and the surface-mount footprint to two different chip select outputs of the microcontroller or address decoding logic, via a pattern of jumper pads that acts like a DPDT switch: The center poles of this "switch" are wired to the chip select lines of the on-board flash device and the proprietary flash socket; the outer poles are wired in a crossover fashion to the "boot" chip select output from the microcontroller (i.e., the chip select line that is asserted when the part is fetching its power-on initialization code or vector table) and to some secondary chip select output. With the "switch" in one position, the on-board flash is selected as the boot device; in the other position, the proprietary module is selected.

With a system like this, you can use the code in your proprietary memory module to write an image to the soldered-down surface-mount flash part. Switch the chip select lines over and you can boot off the surface-mount part and check your code. If it isn't working correctly, you're not locked out of the system and you don't have to desolder the surface-mounted flash part—you can switch the chip selects back over and boot off your proprietary module until you get the final code working.

A similar but considerably more complicated technique with less general applicability is to put a standard header (say, 100 mil DIP) on your board, wired in parallel with the surface-mount pads for the actual chip you intend to use. You can then turn one of the real chips you intend to use into a "cartridge" to fit this header, either by making a small PCB to mount the surface-mount chip, or by using a wire-wrap IC socket as a body to hold the chip (in which case you would wire the chip onto the socket's pins by hand, using wire-wrap wire; this is really only feasible for SOP parts). When you need to reprogram the chip, you simply drop the entire module into the DIP socket of your EPROM burner. The feasibility of this option depends greatly on the capabilities of your EPROM programmer, and your ability to fool it into thinking that the hybrid device you stuck in the DIP socket is in fact the flash chip you have specified, in a surface-mount personality module.

While I'm on the topic of memory, I would like to mention another important rule: Whenever feasible, try to superimpose footprints for different package styles and pinouts on your board, so that you can make running changes to the parts being used. This rule is particularly important for memory devices because for any particular density and technology, there may be several "standard" pinouts. Moreover, if you are a hobbyist, you may find it close to impossible to buy small quantities of specific memory chips, SDRAM in particular. The easiest source for these chips is to buy suitable PC memory modules (SIMMs or DIMMs depending on the vintage of the chips you're trying to find) and scavenge parts off them. Unfortunately, it's usually not possible to know in advance what chips will be on a specific memory module. PC hardware vendors specify their products on density and speed, and they can and do vary PCB layouts and exact chip selections without notice. For this reason alone, it's prudent to design for several different RAM footprints in case you have trouble sourcing a particular chip. However, even without these considerations, it's good practice to lay out for a variety of different RAM and flash pinouts where possible, because prices and availability of these parts can be very volatile.

Another vital rule, particularly relevant to hobbyists and students who will assemble their devices at home, is that you should always be mindful of the assembly process when laying out your board. Compare the two photos shown in Figures 2-2 and 2-3.

Figure 2-2. Cramped layout makes assembly difficult.

The person who designed the board above has placed two large surface-mount components very close to each other. You would encounter some difficulty in hand-assembling this board; no matter which part you fit first, when you come to fit the second part you will have trouble soldering it down because there is not much maneuvering room for a soldering iron. Moreover, the short trace runs between these parts could make probing, patching and otherwise debugging the circuit difficult. Contrast this with the board below:

Figure 2-3. Ample space facilitates hand-assembly.

This board has a generous amount of space between the major components and it would be relatively easy to assemble by hand. (Note that the small passive components don't cause a problem, because we can easily fit those last without needing a lot of room). There is also sufficient room to scratch off solder mask from the traces between the parts and probe, cut or patch them if necessary.

If your board's outline is constrained by an external housing or a need to mate with some other hardware whose size is fixed, consider placing components on both sides so that you can leave sufficient space between them to ease assembly and probing. This is normally something you would avoid at all costs for automated production, since it means the board has to go through the soldering process twice[10], but when constructing by hand it makes no difference to the number of assembly steps.

Speaking of board outlines, it's very valuable to have a PCB vendor who can provide realtime online quotations based on board parameters you enter. The reason for this is that your particular board house will have its own rules about panelization (duplicating your board and fitting it alongside other boards on a single standard-sized plate of blank PCB material) and its own proprietary costing algorithms. The pricing is also usually different for varying tolerances in the production process; for example, a minimum trace/space width of 10 mil will be more expensive than a minimum trace/space width of 25 mil.

I use a PCB manufacturing house called Advanced Circuits (www.4pcb.com) for my prototypes, because, among other desirable qualities, they have a very powerful online quotation system. Once I've finished drawing my schematic, I draw a rough board outline to determine approximately how much area I need to physically fit all the parts. I can then input those parameters to the online quote system and nudge them around to minimize the per-PCB cost. For example, I might modify the board dimensions slightly, or I might decide that I can live without a silkscreen legend on one or both sides of the board if the project is running over budget. I don't sit down and route the board until I've ob-

---

[10] The exact details of this are dependent on the soldering process used by your factory. In general, however, it is cheaper to have components on one side of the board only.

tained at least rough costing and determined the board size, number of layers, trace/hole spacing and other design rules that affect the unit cost.

Also note that if your project contains several small PCBs, it will probably be *much* cheaper for you to prototype if you can draw up an entire set of boards as a single large "master board", with score marks or rows of holes so you can cut apart the sub-boards once you receive them. The same is true if your design requires any board that is non-rectangular in shape; it will be cheaper for you to design a rectangular board and cut it down to size yourself than to have the PCB house put your board on a CNC router to cut it to some arbitrary shape. Some vendors also have special restrictions on particular shapes; for instance, many companies will charge extra for boards that are extremely long and narrow, because these have a tendency to break while being drilled and separated.

The break-even point here depends on several factors, the most important of which is whether your board house charges you extra for large numbers of holes. If you put perforations on your board, you've probably added several hundred drill operations to the production process, which means more time on the CNC drill machine and more wear and tear on the drill bits. Many PCB houses (but not all) will charge for this. Practically all PCB houses will also charge extra to "tab-route" boards (cutting a continuous notch around the parts to be broken away, with small perforated sections holding the parts together during shipping). It so happens that Advanced Circuits charges for tab-routing, but not for drill holes, so I minimize my "master board" costs by demarcating the sub-boards with rows of perforations. I then use shears to separate the sub-boards, and clean up the edges with sandpaper.

Along the same lines, if your project has any surface-mount parts in it at all, you should *never* be tempted to cut costs by ordering your PCBs without solder mask. It is absolutely essential to the assembly process that the board have a mask layer on both component-bearing sides, otherwise you will never be able to clear up all the solder bridges created while populating the board. Similarly, if plated-through holes are optional at your PCB vendor, you should always use them for any non-trivial design, otherwise you will waste far too much time debugging simple continuity issues. It's because of problems like this that I don't recommend that you etch your own PCBs for complex 32-bit projects. Although it can be done, it is practically always a false economy. These high-end projects almost universally require at least two layers (sometimes more—and hand-producing four- or six-layer boards is *extremely* difficult!), and they usually have huge numbers of vias. Hand-drilling these, ensuring continuity and testing such boards is a waste of your time.

As you can see, there are a large number of variables in the design process that will directly affect the per-PCB cost, and if you're on a constrained budget, it would be extremely unwise to commit to a PCB layout without first consulting with the company that will be manufacturing your boards to verify that your design can be made as cheaply as possible.

The final item I'd like to bring to your attention is a reminder that the PCB you're designing is probably based on a largely untested circuit, so you want all the debugging help you can get. If at all possible, try to confine your design to two layers (this will also reduce your prototype PCB costs considerably), or if that's not possible, two layers and two power planes; make every effort to avoid running signals in interior layers where you can't easily probe or patch them[11]. Your life will also be simplified if you run as few signals as possible beneath surface-mount chips and other large parts—keep these traces in the open where you can cut and patch them if you need to. Finally, you may find it very useful to place standard 100 mil or 2mm headers around some of the more complex parts so that you have easy access to all the signals; you don't have to fit the actual headers for production, and they can be very handy for debugging. An excellent example of this is shown in the photograph in Figure 2-4 (this image shows part of the Cirrus Logic EDB7212 evaluation board).

---

[11] This is another good reason to eschew devices in BGA packages. BGA, and especially MBGA, practically demands a six-layer board to achieve easy signal fanout while maintaining solid ground and power planes.

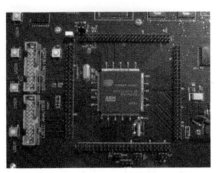

Figure 2-4. Cirrus EP7212 evaluation board headers.

Of course, space constraints on your board may make this kind of luxury impossible, but it is a good feature to include if you can afford the real estate.

When you finally come to order your prototype PCBs, buy *at least* three so that you can make one fully-populated board and have two spare unpopulated boards. You are likely to find that buying three is not much more expensive than buying one; for small production runs, the one-time setup costs of your board dominate the unit price, so while a single board might cost you $100, two boards definitely won't cost $200—they might only cost you $120. (If your product is being laid out and/or manufactured in an outside factory, then you should ask your factory to send you two fully populated boards, in case of accidents, and two unpopulated ones.) Keep one unpopulated board on hand as a reference to use when trying to work out the destination of a trace that disappears under a component. It's much more useful to have a real board for this than just a printout, because you can use a multimeter to do a continuity check and establish beyond any doubt what is connected to what. This technique can also reveal shorts or open circuits that might be hard to locate by looking at a printout or even the Gerber files; it can also reveal errors that are only expressed in the production process—traces that don't quite meet up with pads because of issues with your aperture files, boards that have been flexed and have "popped" their inter-layer connections, pads that have been drilled out too far, and so on.

Always start your debugging of an unproven PCB by bringing up one of the fully populated boards slowly. Put it on a laboratory power supply with the voltage set to 0 and the current limited to a safe value, and gradually turn up the voltage while observing for overheating or overcurrent conditions. This will protect you against obvious problems like power rail shorts, diodes placed on the board backwards, etc. Because you're almost certainly violating rail rise-time specifications for numerous devices on the board, it probably won't start up spontaneously, so once you've brought the input rail(s) up to their nominal supply values without incident, you should apply a hardware reset to the board. You can now verify that the main microcontroller is running; probe the least-significant address bits with your oscilloscope to see that they are toggling (this behavior might be visible only immediately after a reset—if the address bus seems to be frozen, keep watching the least significant few bits while applying a hard reset). Also check that you are seeing activity on the chip select signal for the flash or ROM chip off which your device should be booting—this verifies that any address selection logic between the CPU and the ROM is functional. Once you're sure the microcontroller is getting power and a clock signal and is attempting to fetch code from ROM, you can try to load some firmware onto the board and begin testing the remaining hardware.

The reason I suggested that you should order two unpopulated boards is in case you can't get the fully populated board to boot up at all. In such cases, a useful technique is to build up one of the boards yourself by hand, testing continuously. Start by adding power regulation and conditioning components; check the rails to ensure that these regulators are providing the correct output voltages. Then fit only the main microcontroller and any external passive components (crystal or ceramic resonator, RC network, oscillator module, etc.) it requires to start operating. Bring up the board again and observe the address bus and chip select lines. If you see activity on these immediately after a reset, you should add the bare minimum of hardware required to be able to load firmware onto the board. Once you are able to run your own code on the board, you can add peripherals one by one until you find the problem that was originally preventing the board from starting correctly.

If all the above sounds like a flippant summary of a painstaking, days-long process—well, unfortunately it sometimes is. My experience, however, has been that first-round boards which are a total "no-go" on delivery are usually suffering from a gross problem that can be identified early on in the debugging process. The subtle problems generally arrive much later in the development process, when your firmware is largely complete and you are giving the hardware a more thorough workout. Basing your circuit on an existing reference design will also shorten the debugging process considerably.

Hints for Surface-Mounting by Hand

If you choose to develop on your own custom board, you will probably need to assemble at least one or two units by hand. Fortunately, prototype construction with surface-mount parts is not as difficult as you might think, even if your funds don't stretch to professional hot-air rework equipment. Provided you don't intend to assemble more than a couple of units, you can fairly easily mount parts all the way down to fine-pitch QFP (quad flat pack) using nothing more exotic than a normal temperature-controlled soldering iron, tweezers and desoldering braid. In all cases, you should be working under excellent light, preferably with magnification. A bench-mounted illuminated magnifier with a toroidal fluorescent lamp and jointed, swiveling arm is ideal.

The techniques you will use for surface-mount assembly are very different from those used with through-hole parts. The most obvious difference is that you usually do not first place the component then solder it. Because most surface-mount parts have no way of remaining in alignment until soldered down, and because you have only two hands, you need to apply solder to at least one pad and place the component on the wet solder bead to stick it down. There are three major methods you will use, depending on the type of component being mounted:

- **For parts with two pads or leads**, such as surface-mount capacitors and resistors, begin by putting a solder bead on one pad of the PCB. Keep your soldering iron on the pad so the solder stays melted, and use tweezers to place the part on the board as accurately as possible. Remove the soldering iron, and continue to hold down the part until the solder has cooled. Remove the tweezers carefully; the component will be held in position by the end you soldered. Now solder down the other end.

- **For parts with a few leads**, such as voltage regulators and transistors, or ICs with relatively coarse-pitch leads (regular "gull-wing" SOIC packages, SOJ and so on), use a similar technique to that described above. Put solder on one pad on the PCB, stick the component down on that one pad while the joint is still hot, and use fine solder to solder down the remaining leads.

- **For parts with many, fine-pitched leads** (QFP, TSOP and similar packages), first ensure that you have the PCB lying flat on a solid surface. Place the component down onto the PCB; orient it over its lands as accurately as you can. Press down firmly to keep the component oriented correctly while you tack down the corner pins. Don't worry if you create bridges between pins; just dab enough solder onto pins around the device so that it will hold itself down in alignment with its pads. Once you have the device secured, go all around it wetting all the joints with a liberal amount of solder. **Work as quickly as you can on this step**, and allow the device time to cool during the process. (Avoid the use of spray-on component coolers though, especially on ceramic devices, or you may crack the package). Once every joint is liberally covered in solder, use desoldering braid to remove all the bridges. Don't worry that you are "desoldering" the component—enough solder will remain to keep the pins bonded down firmly. Again, you should avoid applying heat to the device for extended periods of time; do a few pins, then allow the device to cool down before attempting more pins.

There are several different grades of desoldering braid, and most of the types I have tried have not been very good for this type of work. The specific product I use, with excellent results, is "One-Step Braid" from Easy Braid, Minneapolis, MN, catalog #OS-A-25; readily available from major online component dealers like Digi-Key. This is a very thin braid composed of fine copper filaments impregnated with resin-type flux. When you apply it to a hot joint, it wicks the solder away very quickly. Unfluxed braids don't draw away the wet solder quickly enough. Wider braids are hard to use accurately; they also act as splendid heatsinks, so with a normal soldering iron of moderate power it can be hard to remove the wick once the joint has been adequately cleaned. In attempting to reheat the used braid to remove it, you can easily create more solder bridges; very frustrating, to say the least.

Note that all of these soldering techniques carry an inherent danger of overheating the components and PCBs. They are very much in violation of manufacturer-recommended soldering stresses and are **suitable for prototype work only**. It's also essential that you practice before attempting to work on a real project. You can obtain prebuilt PCBs with several surface-mount land patterns etched on them from most electronics supply houses, but these PCBs are rather costly. (They are sold for use by repair technicians being trained in surface mount rework techniques.)

A much cheaper way to acquire sacrificial practice boards is to rescue some elderly computer hardware or consumer electronics from your junkpile, but if you choose this route you will first have to remove the chips before you can try to solder them back on. To achieve this, I normally preheat the board using a hairdryer, then use a small handheld butane torch—about the size of a disposable cigarette lighter, and readily available at electronics parts stores and mail-order houses—applied in a rapid circular motion around the part to be removed. Once the joints on all sides of the part are melted, quickly flip the board upside-down and give it a sharp tap to break the surface tension and remove the part. I recommend using a shoebox full of crumpled tissue paper to catch the falling components. By the way, **never attempt to remove parts from phenolic PCBs this way**; the board will catch fire and emit poisonous, evil-smelling smoke. In the present day, you are most unlikely to receive phenolic material from a commercial board vendor, but there is still a lot of coppered phenol board in the hobbyist market, so it is a warning worth noting if you etch your own boards.

These simple techniques will work surprisingly well for almost every common IC package short of BGA (and a few other completely leadless packages such as MLF and TCP). There isn't any truly reliable, inexpensive way to hand-prototype BGA parts with common workshop equipment; even on a real production line with industrial-grade assembly tools and side-looking X-ray equipment for diagnostics, there can be problems. I have attempted to hand-assemble these devices on a few occasions, and my final verdict is that the failure rate for hand-assembly is too high to make it cost-effective. Sockets do exist for adapting BGA to PGA, which you can use on a prototype easily enough – but these sockets are astonishingly expensive (hundreds of dollars apiece). If you absolutely have to use a BGA device, you may be able to inveigle your chip vendor into supplying one or two chips mounted on PGA adapter boards for prototyping. If not, the simplest route is to use a contract assembler to place the BGA devices on your board for you.

Choosing PCB Layout Software

Regardless of how you intend to prototype your device, unless you are using an off-the-shelf single-board computer you will almost certainly need to lay out your own PCBs. Even if you do use an off-the-shelf computer, your life will be made easier if you can make a PCB to hold any "glue" circuits that need to be attached to that computer.

CAD software for creating PCBs is likely to be the single largest tool expenditure you will need to make, depending on your needs, and the tool you select can make a huge difference in development time, particularly if you are working with an external factory. Roughly speaking, PCB CAD packages can be divided into three categories:

- "Mainstream" products such as Easytrax, PADS, OrCAD et al.

- Less widely-known commercial packages such as Cadsoft EAGLE, shareware packages, etc. These packages are considerably cheaper than mainstream products, but may have limitations. These packages can generate industry-standard artwork and CNC drill control files that can be sent to any PCB fabrication house.

- Completely proprietary packages geared to a single PCB fabricator. These are distinguished from the preceding category by the fact that they cannot generate industry-standard output file formats such as RS-274X Gerbers or Excellon drill control files. These packages are usually free.

There are times when it is unambiguously a sound investment to use a mainstream package. If you intend to commercialize your product, and you don't intend to handle every aspect of the actual manufacturing, it makes good sense to use the same tools your factory will be using. Even in large corporations, it's quite common for the circuit to be designed by head office in the U.S. or Europe but the final PCB layout to be developed by a contract manufacturer, typically in Asia. This arrangement allows the people on the front line in the factory to tweak the layout for housing mold changes, airflow, ease of final assembly, compatibility with factory test fixtures and other minutiae, easing the workload on you.

Of course it's entirely possible for you to design the circuit and prototype PCBs in your preferred software package, then give printed schematics and Gerbers for your prototype to the factory for transliteration into their favored package. However, I strongly discourage you from this approach, because it introduces a high probability of errors. Although there have been some promising interoperability noises made by a few of the major CAD vendors, at the moment there is no standard data format for schematics and PCB metadata (keepouts, track width and restring constraints, autorouter parameters, part autoplacement rules and so on). Third-party conversion utilities for some formats do exist, but they don't always do a perfect job, and not all formats are supported. In the worst case, a PCB layout engineer in your factory will sit down with a printout of your schematic and enter it by hand into their software; for any non-trivial circuit, this process is guaranteed to result in some errors which you have to find. Furthermore, unless the factory spends a lot of extra time generating additional metadata, it won't be possible for them to do any automated electrical testing on the boards they produce. Delays from the factory, its vendors, shipping and customs clearance are often three to four weeks; allow a similar amount of time to test the board and locate any layout bugs, and your project could easily be delayed for as much as two months for every board respin. If you absolutely HAVE to do things this way, then please refer to the board bring-up techniques discussed in a later chapter.

If format translation problems aren't a consideration for you (for example if you intend to build the product yourself without involving an external factory who might need to alter the layout), or you're working on a budget and don't want to pay the price of one of the big-name packages, then your best option is one of the less expensive commercial packages. There are a large number of low-cost commercial, shareware and even freeware PCB CAD packages, but my personal budget-conscious recommendation for CAD software is EAGLE from Cadsoft; the freeware versions of this package for both Windows and Linux are included on the CD-ROM in the /eagle directory. This version is free for non-commercial use (please refer to the license agreement for more information), and it has all the libraries and features of the full version—including autorouter support—but it is limited to two PCB layers and a maximum board size of 100x80mm. The full product supports arbitrary size multilayer boards. EAGLE has numerous advantages which lead me to use and recommend it:

- It's reasonably priced (at the time of writing, licenses start at $49—unconstrained commercial licenses are about a quarter of the price of similar mainstream packages).

- The free demonstration version is fully usable for moderately complex projects.

- It offers side-by-side schematic capture and PCB design. I consider this feature essential in a PCB CAD package; entering schematics in one piece of software and generating PCBs in another is a fast route to expensive transcription errors.

- An autorouter and fairly comprehensive starter libraries (including numerous surface-mount parts, and pre-drawn footprints that you can re-use for your own parts) are included.

- It is available for both Windows and Linux, which is helpful if you have only one PC and you are developing your firmware under Linux; you don't have to restart your PC every time you want to modify the schematic.

- EAGLE has a sufficiently large userbase to have reached the "critical mass" for peer support. Cadsoft runs a publicly-accessible NNTP server carrying several categories of discussions; this userbase is active and very helpful. Being able to converse with fellow users of a complex piece of software is invaluable, especially when first getting up and running with the product.

- The product doesn't employ infuriating hardware-based copy-protection schemes.

There are obviously some downsides to using non-mainstream PCB CAD software, besides the interoperability issues mentioned above, and you should be aware of these. Firstly, component vendors and third-party CAD librarians will mostly not support your particular CAD package; this will involve you in some additional maintenance work building footprint libraries for your parts. Secondly, and perhaps more importantly, complex features like autorouting and autoplacement are not provided in most of these packages. (EAGLE does include an autorouter, but it lacks autoplacement, and at the time of writing it does not support blind vias, meaning that it cannot easily be used to develop PCBs containing large BGA-package devices. The autorouter is also rather primitive; on complex boards it is not unusual have to run several route operations, nudging components this way and that, before the board will route fully). Finally, none of the low-end CAD packages I have evaluated directly support linkage to external programs for product casing codesign, thermal/airflow simulations, SPICE simulations, generation of 3D mechanical models, etc. Keeping these facts in mind, you need to analyze your requirements carefully before deciding where to direct your money.

As far as the third category of PCB CAD software goes, I won't mention any of these by name, because I see few or no benefits, and huge disadvantages, in using this type of software. Briefly, certain online PCB fabrication houses supply PCB layout programs for free download. The problem is that these programs use their own proprietary storage format, and usually can't directly generate industry-standard Gerber and drill control files. The idea is that you use the free software to lay out your board and send it to the fabrication house that supplied the software; they use another tool (which they don't make generally available) to convert the proprietary file into something their CAM systems can use. Besides the obvious fact that using this software locks you into using that particular fabrication house (which may not be the best-priced, or may have delivery problems), it also means that your precious PCB data is hidden inside data files that can't readily be exported or converted. When your needs change (e.g. once you start mass-production) or if your chosen fabrication house goes out of business, you might be stuck with re-engineering the board all over again.

## Chapter 3

The GNU Toolchain

Building the Toolchain

In this section, we'll go over the steps required to build the various components of the GNU toolchain for C, C++ and assembly-language programming on one specific embedded target (in this case, ARM). This is all strictly utilitarian "how-to" information, more of an installation guide than descriptive text. For details about what the parts of this toolchain do, please refer to the next section.

If you intend to run the tools under Windows, your first step will be to install Cygwin (If you're using Linux, skip this step). Cygwin version 1.3.16-1 is included on the CD-ROM in the "cygwin" directory. Double-click setup.exe, click "Next", choose "Install from Local Directory", click Next twice, and you will be shown the Select Packages dialog.

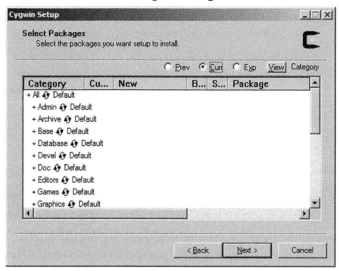

Figure 3-1. Cygwin package selection dialog.

An important note applies here: The default setup configuration for current versions of Cygwin will not download or install native development tools. These tools—Cygwin-hosted, Cygwin-targeted versions of gcc, gas, ld et al – are required in order to build your cross-compiling tools. In the Select Packages dialog shown above, you must click on the word "Default" on the line headed "Devel". There will be a long pause, and the word "Default" will change to "Install." This enables installation of the required packages. If you would like to have native text editor support in Cygwin, then perform the same step on the line headed "Editors."

Once you have made these install change(s), click Next and wait for the product installation to complete. Click "Finish" in the final dialog and you're done.

Tip: The CD-ROM with this book contains a complete set of install files for all possible setup options, so you should not need to download any additional modules from the Internet. However, if you go to the Cygwin web site for a newer version of the product, you are likely to run into an irritating bug in the online setup program. This bug appears when you are installing off the Internet, and you need to enable or disable a large number of items in the "Select Packages" dialog shown above. If you spend too long in this dialog—and selecting or deselecting a complicated option can take a couple of minutes—the FTP connection opened by Setup will expire. The install process will appear to complete very quickly, but nothing will actually be downloaded and you will get an error message asking if you want to retry. Select "Yes", and you will be returned to the dialog that allows you to select an FTP download server. Select a *different* server from the one you selected when you first ran the setup program, and a new FTP connection will be established and the install will proceed normally.

The core of Cygwin—and the component to which the "Cygwin version number" refers—is a Windows DLL that provides a sort of UNIX emulation layer; a pseudo-operating system that translates many standard UNIX APIs into Windows APIs. Many UNIX programs can be cross-compiled to run on Cygwin, and the distribution that is on the CD-ROM actually consists of a large number of separate utilities, precompiled for Cygwin. For example, the icon that appears on your desktop for Cygwin is actually a link to the "ash" shell, not Cygwin per se.

You can check the installed Cygwin DLL version at any time by opening a Cygwin shell prompt and entering the command:

```
uname -r
```

With Cygwin installed, for those users that need it, we're ready to start building the actual cross-compiler and other tools (ensure that you have about 550Mb of free space on the hard drive that holds your /tmp directory; the temporary files can be large). Start by copying the .tar.gz files into your /tmp directory. If you're running Linux, simply mount the CD-ROM and `cp /mnt/cdrom/gcctools/* /tmp` to achieve this.

If you're using Cygwin, the easiest way to "import" the files is simply to drag and drop them using Windows Explorer; if you installed cygwin to the default directory of c:\cygwin, then this will be c:\cygwin\tmp. Once you've copied the files over, double-click the Cygwin icon on your desktop to open a Cygwin session. From here on, build instructions are identical for both Windows and Linux.

We begin by uncompressing the various modules:

```
cd /tmp

tar zxvf binutils-2.13.1.tar.gz

tar zxvf gcc-3.2.tar.gz

tar zxvf gdb-5.2.tar.gz

tar zxvf newlib-1.10.0.tar.gz
```

At this point you can delete the tarballs (`rm -f /tmp/*.gz`). Now we configure and build binutils, which is a simple process that rarely causes problems:

```
cd /tmp/binutils-2.13.1

./configure --target=arm-elf --prefix=/tools/arm-elf

make all install
```

The --prefix switch sets the location for all our binaries to be installed; the --target switch sets the type of processor we're building for, the build environment, and the type of executable. arm-elf is a generic ARM target, without OS-specific modifications, that generates object files in the ELF file format. Some other targets that may be of interest are as follows (descriptions are suggested applications only):

- arm-linux – Linux ARM targets.

- i386-pe – Generic Intel x86 code for embedded PC platforms.

- m68k-elf – Motorola 680x0 embedded targets.

- m68k-coff – Similar to m68k-unknown-elf, but you might use this configuration if you need to link against COFF formatted libraries. COFF is older than ELF and its use is deprecated. Some code, for instance eCos, cannot be built using COFF object file formats.

- mips-elf – MIPS embedded targets.

- sh-elf – 32-bit Hitachi SuperH embedded targets.

- xscale-elf – Intel XScale embedded targets.

There is currently no official master list of the target names that are supported, but you can glean some information about supported targets by referring to the configure.in file in the binutils source directory; look for the case statement at line 271.

Before continuing, we need to add our ARM tools to the PATH, because subsequent build steps require them[12].

```
PATH=/tools/arm-elf/bin:$PATH
```

Now we're ready to build gcc. This is unfortunately not a one-step operation[13].

---

[12] Unless you want to set the path manually every time you open Cygwin or reboot your Linux machine, you'll need to add these tools to your path permanently. This is accomplished by editing the startup script for your shell. For example, on a Linux system running bash, adding the PATH= line to the file ~/.bashrc will ensure that the path is set correctly every time you open a shell.

[13] The install instructions for eCos describe a single-step build process for gcc. The compiler that results from this process is incomplete, and cannot be used to build general-purpose programs. The instructions given in this book build a complete compiler that can be used to build eCos or standalone programs equally well.

```
cd /tmp/gcc-3.2
```

```
./configure --target=arm-elf --prefix=/tools/
arm-elf --enable-languages=c,c++ --without-
headers --with-newlib
```

```
make all-gcc install-gcc
```

At this point, we have a *partially* usable cross-compiler for ARM code; it's functional enough only to enable us to build our selected run-time library and bootstrap our way to building the rest of the compiler. In order to build a fully-functional compiler, we need to build and install a standard library such as newlib or glibc. We are using newlib:

```
mkdir /tmp/newlib-build
```

```
cd /tmp/newlib-build
```

```
../newlib-1.10.0/configure --target=arm-elf
--prefix=/tools/arm-elf
```

```
make all install
```

Important: Some references on the Internet will direct you to simply unpack and configure newlib without making a separate temporary build directory. I suspect that these instructions are either cut and pasted from some very old original source (written at a time when this method actually worked), or the authors have assumed that it "ought to work" without actually testing it. With current versions of newlib, at least under all the host operating systems I have tested, it is impossible to build the library correctly in the sources directory; you MUST create a temporary build directory as shown above.

We now have all the components required to build a fully working C/C++ compiler.

```
cd /tmp/gcc-3.2
```

```
./configure --target=arm-elf --prefix=/tools/
arm-elf --enable-languages=c,c++ --with-newlib
```

```
make all-gcc install-gcc
```

The last item we need to build is gdb, the GNU debugger. This is accomplished easily enough:

```
cd /tmp/gdb-5.2

./configure --target=arm-elf --prefix=/tools/
arm-elf

make all install
```

Note that gdb 5.0 was the current, recommended version for a comparatively long period of time, so many references still mention it; for example, the install instructions for eCos still specify this version. There are some annoying issues with gdb 5.0 when using the ARM remote debugging protocol (for instance, when communicating with ARM's Angel ROM monitor), so I recommend using the more recent version, unless there is a specific reason why you must use some other version.

One final note: If you are using the Macraigor Wiggler JTAG debugging pod mentioned earlier (or another of Macraigor's debugger modules), Macraigor Systems makes your life easy by providing precompiled versions of binutils, gcc and gdb (for Cygwin, Linux and Solaris hosts) for various microcontroller cores supported by their hardware. To obtain this software, visit their homepage at www.ocdemon.com. (You don't actually need to own Macgraigor's hardware to use these precompiled tools; they are generic.) Even so, it is important that you know how to build the components of your toolchain, because it might be expedient or necessary for you to use a different combination of tool versions, or to choose customized build options for your particular circumstances.

Overview of the GNU Build Environment

The GNU toolchain referenced in this book, when used for embedded development, consists of three major modules, each of which is composed of several sub-components. The major modules are *binutils* (a collection of miscellaneous underlying utilities including an assembler, a linker and so on), *gcc* (the GNU C/C++ compiler[14]), and *gdb*, the GNU debugger. The CD-ROM

---

[14] Gcc is actually much more than just a C/C++ compiler; it has support for other languages (at the time of writing, Objective-C, Fortran, Java and Ada). We will only discuss C and C++ in this book, partly for expediency, but mostly because these languages are most frequently used for embedded applications.

included with this book includes binutils version X, gcc version 3.2, and gdb version 5.2.

In addition to the modules mentioned above, you will almost certainly want a C run-time library. The run-time library you choose depends on your underlying operating system; it provides a variety of OS-independent functions such as string manipulation, memory copy and compare functions, and so forth, as well as standardized interfaces between your code and operating system features such as filesystems. If you are developing for Linux, you will probably use the GNU C library, glibc. If you're developing for eCos, you don't need a separate standard library, because all the functions you need are part of the eCos operating system library against which you link your application. For systems that don't have an operating system, probably the most popular choice is newlib, which is small and OS-agnostic. It provides a large set of handy functions at a relatively low memory cost. Since our example project is a standalone device, we're going to use newlib, and so I have included a recent version on the CD-ROM.

Note that all these component versions are not necessarily the latest available, although they happen to be the latest release versions at the time of writing. The reason I selected these specific versions is that I know they work well for ARM7 targets, which is the platform I'm principally discussing, and that they build under Linux and Cygwin with approximately equal steps and no special caveats. This neatly leads me to an important warning about using this kind of extremely general-purpose tool: the GNU toolchain is a very large project maintained by a large number of independent developers, and it is constantly evolving. Since not all developers are interested in the same enhancements, and due to the ENORMOUS resources required to perform full testing of all the possible host and target options this toolchain supports, it is not uncommon for incremental changes in various parts of the toolchain to break (temporarily) support for particular targets and/or hosts. The problems that will result range from obvious and catastrophic (e.g., you can't compile the toolchain at all) to subtle (e.g., a change in stack handling that makes code from the new toolchain partially incompatible with code generated by older versions). The subtle problems are by far the worst, because they don't necessarily cause any errors or warnings when you're compiling either the toolchain or your own code; they silently insert problems that lurk in your object code waiting for inopportune moments to make themselves visible.

Besides these unintentional issues, from time to time certain unpopular targets or hosts simply lose momentum and their branch is either cut, or eventually atrophies off the source tree. For example, the Hewlett-Packard PA-RISC core found in several low-cost microcontrollers (intended for Internet appliances) is no longer supported by the GNU toolchain. Support for these parts hasn't been actively removed, though; it's just not being updated alongside the more popular code. In practical terms, this means that the current versions of binutils, gdb and newlib will build properly for PA-RISC targets (but won't necessarily work properly!), but the current version of gcc will not build at all when configured for the PA-RISC microcontroller parts for which I have tested it.

In order to avoid all these sorts of problems, the best policy is to pick a specific set of GNU tools that are known to work on your chosen host and generate known-good code for your target platform. If your chip or OS vendor supplies or recommends a specific version, then use that version unless you have an inescapable need for a feature that is only found in newer versions. If you need to research this yourself, one excellent way of establishing which versions to use is to look at a large, well-established project (e.g., a Linux or NetBSD port for your processor core) and determine what versions the maintainers of that project currently recommend. Failing that, you should pick the most recent official release, as opposed to pre-release, and work with that.

The object of the discussion above is by no means to cause alarm, but simply to dissuade you from downloading the latest version of everything and expecting it to work perfectly under all circumstances. In particular, I want to discourage you from "upgrading" components of your toolchain to the latest version just because a new version becomes available. The GNU toolchain is almost certainly the fastest-evolving piece of development software on the planet, which is fantastic news for people working with new CPUs, because it means they can expect free tools to be available shortly after the part hits the market. The flipside of that fact is that the "latest" version of these tools is usually a work-in-progress intermediate build. From time to time, there is a "stable" release, which then becomes the latest "official" version of gcc. This is comparable to the point at which a commercial compiler would issue a new version.

In any case, whether you followed the instructions in the previous section or obtained a ready-made toolchain from an outside source, you now have a working set of GNU tools for your target processor. Let's discuss some of the utilities that comprise this toolchain. Note that the versions you have installed on your system all have the "arm-elf-" target name prepended to their names. For instance, to run as, you would use the command arm-elf-as. (The only exception to this rule is make, which is invoked simply as "make".) For information beyond the descriptions below, you should refer to the on-line documentation, which can be found in three primary locations:

1. Command-line help, which can be viewed with the --help switch—for example, `arm-elf-gcc —help`. This help is extremely brief and is mostly limited to a description of command-line switches.

2. Info pages, accessible with the info utility—for example, `info gcc`. This documentation is detailed, and is presented in an easily-browsed hypertext format.

3. HTML conversions of the info pages, available at the GNU web site (www.gnu.org). There is nothing here that you haven't already got in the info pages, but some people prefer to browse the documentation online with their web browser.

By the way, there are no pages missing from your book: it's true that one major program I'm not explicitly documenting here is gcc itself. The reason for this is that gcc is a standards-based compiler, and all you really need to know in order to use it is how to program in C. The bulk of what you absolutely need to learn in order to use gcc is not really about the compiler itself, but rather the infrastructure that supports it (newlib, gas, ld, and so on), and this information is presented below in some detail.

There is a small amount of additional help you might need on a project-by-project basis, such as gcc-specific #pragma information, command-line switches, and so forth. There is nothing here that really requires elucidation over and above the gcc online documentation, so rather than simply cutting and pasting that documentation into this book, I encourage you to refer to the info page for gcc.

GNU Make and an Introduction to Makefiles

Make is the puppetmaster of your build environment; it checks source file dependencies, compiles only those files that require recompilation, and executes whatever other commands are required to build your application. Although it is possible to build programs without learning to use make, you will find that this utility makes your life much easier and development considerably more efficient. If you're used to programming in an integrated development environment (IDE) like Microsoft Visual C++, you can think of the makefile as your project file.

If you run make with no command-line arguments, it will look for a file named Makefile in the current directory. (Actually, it will search for several different possible default makefiles by name, and you can override this behavior to choose some other file as your makefile, but consistency and aesthetic reasons make it a good idea to call your makefile, Makefile). This is a simple text file, with the following general format:

```
target : prerequisites
        commands
        {... possibly more than one command}
```

Important: The *command* line(s) must start with a hard tab character! This requirement frequently trips up programmers new to makefile syntax; if you use spaces (soft tabs) instead, make will report errors that don't directly seem to relate to the problem. Ensure that you use a text editor that preserves hard tabs when it saves files to disk.

When you run make, if you don't specify a specific target to build, it will look for the first target whose name doesn't begin with a period, and attempt to build it. Make looks at the list of prerequisite files, checks their dates and determines what, if anything, needs to be rebuilt in order to build the target. The remainder of the makefile may contain instructions for building these prerequisite files, and their sub-dependencies (if any).

For example, let's consider a simple program that consists of two C files (main.c and functions.c) and a single header file (mydefs.h) that is included by both C source files. A simplistic makefile for this program might look like this:

```
myprog: main.o functions.o
   arm-elf-gcc -o myprog main.o functions.o

main.o: main.c mydefs.h
   arm-elf-gcc -c main.c

functions.o: functions.c mydefs.h
   arm-elf-gcc -c functions.c
```

When you invoke make for the first time, it will first encounter the target you're trying to build—myprog—and begin examining its prerequisite list. Since this is the first time you've tried to build the program, the two .o (object) files will not yet exist, and so make will look further in the file for instructions on how to build them.

In its quest to build myprog, make will first try to build main.o. It sees that the files main.c and mydefs.h exist on the hard disk, and they are newer than the (nonexistent) file main.o. It therefore executes the command `arm-elf-gcc -c main.c` to build the object file. Likewise, functions.o is built. Note that the prerequisites to a target are built in the order they appear on the prerequisite list, *not* the order in which their target rules appear in the makefile. Also note that if there is some kind of error in the source file, arm-elf-gcc will return an error status to make, and the entire build operation will stop.

Tip: Often, a lot of output can be generated before the build process stops, and you may want to capture that output for debugging purposes. You can redirect the output of make using the redirect operator >, for example `make > make.out`. However, this will only redirect the stdout (standard output) channel, not stderr (the error channel), and some programs always emit their error messages to stderr; furthermore, it suppresses output to the screen, so you can't see what's going on during the build process. For this reason, you will often see people invoke make with a command like this: make 2>&1 | tee make.out. This directs both stdout and stderr to both the screen and a log file.

So much for the first run building our simple program; you may be thinking that the same functionality could easily be achieved by writing a shell script that compiles main.c, functions.c and links them. (In fact, this would be a single command.) However, let's consider what happens when you change just a few files out of a (hypothetically large) project. Suppose, for instance, that you make a change to main.c. When you next run make, it checks the dependencies for main.o and sees that main.c is newer than main.o. Therefore, make calls gcc to build an updated version of main.o. Functions.o, on the other hand, is still up to date, and doesn't need to be built. On the other hand, if you change mydefs.h, make will see that both main.o and functions.o need to be rebuilt. All this automatic functionality makes your life as a programmer much easier.

If you wanted to, you could also build just part of our program by invoking make with the name of the target you want built - for example make main.o. This functionality isn't terribly useful for our simple example, but for more complex projects that contain numerous sub-targets, it's very useful to be able to confine your build attempt to a single target. This is particularly useful when you're porting code from one operating system or processor to another; restricting yourself to a single target cuts down the number of error messages you have to see and analyze. Once you've got that single target building correctly, you can move on to the next target.

Having said all this, note that make is a "dumb" tool—it works by looking at the timestamps on files and the error exit codes reported by the programs you ask it to invoke. As a result, there are dependencies it will not catch, such as a line in your program like this:

```
printf("This program was built on " __DATE__ "\n");
```

__DATE__ is an internal macro built into gcc, which evaluates to a string describing the current date (when you built the program); sometimes it's useful to show this information to the user. However, since make doesn't know that your sourcefile references this volatile piece of information, it won't rebuild that source file automatically. You can easily force make to consider certain files as infinitely new (i.e., always need to be rebuilt) using the --assume-new switch, e.g. `--assume-new=main.c`, but this might mean that substantial portions of your program need to be built every time you change anything, thereby negating much of the benefit of incremental compilation and linking. It's usually easier, at least for cosmetic items like a build date string, simply to allow that one module to become inconsistent.

In order to make sure that we resolve all these kinds of issues before making a final shippable build of our program, it would be nice to clean up all temporary output files and let make do a complete build from scratch. That way we can be sure that the resulting binary file contains the latest revisions of everything and all dates and other such information are consistent. We can achieve this simply by deleting all the output files that can possibly be generated by a make run, and then invoking make to rebuild our target again. In the case of our simple program, we generate three files: myprog, main.o and functions.o, so we can clean up with one simple command:

```
rm -f myprog main.o functions.o
```

(The -f switch tells rm not to report an error if called upon to delete a nonexistent file. Make also has a general-purpose syntax for ignoring errors on specific build steps—you simply prepend a minus sign to the command line in the makefile—but adding the -f switch also keeps the make output tidy by preventing rm from putting an error message onscreen.)

It would be more useful, though, to be able to embed this behavior into our makefile, so that we don't have to remember the cleanup steps for a particular project. We can achieve this easily by adding the following stanza to our makefile:

```
clean:
        rm -f myprog main.o functions.o
```

Invoking `make clean` will remove our object files and ready the build environment for a from-scratch build. As you can see from this example, a target doesn't have to be the name of a file that needs to be built; it can simply be a label you give to a specific sequence of commands.

Even in the above simple makefile example, there is some duplicated typing of filenames in essentially identical stanzas. Make has several features that allow you to cut down on this typing. The first of these is the ability to set and read string variables:

```
OBJS = main.o \
       functions.o
```

OBJS is one of several commonly used names for the list of object files required to build a project. (Other common names are objects, OBJECTS, obj and OBJ.) Note the backslash at the end of the first line above; similar to C syntax, this conjoins the first line syntactically to the line below. Putting each object file on its own line allows for easier cutting and pasting.

Our model makefile now becomes:

```
OBJS = main.o \
       functions.o

myprog: $(OBJS)
        arm-elf-gcc -o myprog $(OBJS)

main.o: main.c mydefs.h
        arm-elf-gcc -c main.c

functions.o: functions.c mydefs.h
        arm-elf-gcc -c functions.c

clean:
        rm -f myprog $(OBJS)
```

At every place in this makefile where $(OBJS) appears, make will substitute the value of OBJS; i.e. "main.o functions.o".

We still have quite a bit of redundant typing, though, and it will get much worse as our program grows more and more modules. Fortunately, make supports a number of rules for common functions such as turning a .c file into an .o file (the default behavior is usually "cc -c file.c -o file.o"; these built-in rules are termed *implicit rules* in make jargon). You can learn about the hardcoded implicit rules from the info page for make. These rules are, however, not usually correct for embedded targets, because we want to run the cross-compiling versions of utilities such as gcc, and make will by default try to run the native versions[15]. We will build our own set of rules as necessary using *pattern rules*.

Pattern rules are general instructions for building a file of type *x* from a file of type *y*. For example, we can use the following rule for compiling our C source files:

```
%.o : %.c
        arm-elf-gcc -c $(CFLAGS) $< -o $@
```

This rule says "In order to build a *something*.o file from a corresponding *something*.c file, perform the command specified on the second line." $< and $@ are special placeholders for the name of the source (.c) and target (.o) files, respectively. These special strings are called *automatic variables*. There are quite a few other automatic variables provided for you by make, but they aren't necessary for building the example code in this book. For more information, refer to the make info page.

Note that CFLAGS, which you see mentioned above, is another commonly found variable used to represent whatever command-line switches are required to build our code. For example, we might want to specify the gcc optimization switch -O3. By putting these switches in a variable, we can define them once at the start of the makefile and easily make global changes.

If we're using pattern rules, we no longer need to list the C sourcefile as a prerequisite for each module; we only need to list the dependency files that make can't infer by looking at directory information. This slims down our makefile somewhat:

---

[15] It is possible to get make's hardcoded implicit rules to work for cross-compilation by tinkering with aliases or by using different settings for environment variables. However, to my mind this makes the development environment more complicated, especially if you routinely work with more than one target processor and/or different host operating systems. The fewer assumptions you make about the build environment, the easier it will be to write a portable makefile. This is particularly important if for some reason you need to use a non-GNU version of make.

```
CFLAGS =
OBJS =  main.o \
        functions.o

myprog: $(OBJS)
        arm-elf-gcc -o myprog $(OBJS)

%.o : %.c
        arm-elf-gcc -c $(CFLAGS) $< -o $@

main.o: mydefs.h
functions.o: mydefs.h
```

(If you don't care about checking dependencies against the header files—and it can be something of a chore to maintain this information by hand—you could omit the last two lines. Just remember that if you take this shortcut, then change a header file, you will have to make clean and then make to be sure that your changes are propagated through the entire program. Since I find that most of the incremental changes I make and test are alterations to the sourcecode rather than header files, I personally tend to use totally generic makefiles that don't check header dependencies.)

Now let's suppose that we need to add a small assembly-language file called boot.s to our program. Although one doesn't often need to do this when writing a general-purpose program, it's normal in embedded applications to need a little assembly-language glue to handle power-on startup, interrupt vectors and so on. To do this, we'll add a new implicit rule to handle .s files[16], and a new variable, $(ASFLAGS), to store any command-line switches we might want to pass to the assembler. We'll also add boot.o to the list of prerequisites for myprog. The new makefile looks like this:

```
CFLAGS =
ASFLAGS =

OBJS =  boot.o \
        main.o \
        functions.o
```

---

[16] The default extension for assembler source files on non-Intel platforms tends to be .s. If you're more comfortable with it, you can use .asm instead, of course.

```
myprog: $(OBJS)
        arm-elf-gcc -o myprog $(OBJS)

%.o : %.c
        arm-elf-gcc -c $(CFLAGS) $< -o $@

%.o : %.s
        arm-elf-as $(ASFLAGS) $< -o $@

main.o: mydefs.h
functions.o: mydefs.h
```

Tip: ld (which is invoked by gcc to link the executable; ld is discussed in detail below) will link files in the order they're provided on the command line. This piece of information can be important, because it affects where your code winds up in the final image. There are frequently special reasons why you might want particular pieces of code to be close to each other in memory—for example, so that you can use short-form relative addressing modes for time-critical interrupt handlers. In this case, I've put boot.o at the start of the link list (i.e., the lowest memory address) because in the ARM architecture, power-on reset and interrupt conditions cause jumps into a table at location 0x00000000 in memory, and this part of our code will need to be hand-crafted in assembly language.

Gas—The GNU Assembler

As, or rather gas, is the GNU assembler. In general, the GNU authors keep their assembler syntax fairly close to the manufacturer's published mnemonic conventions. Nevertheless, there are some differences between gas and the assemblers built into other toolchains that will cause you issues when porting code (for example, from a chip vendor's application notes). In most cases, these problems arise because the GNU arm-elf assembler intentionally matches syntax with other targets for the assembler; in some cases, there are simply vendor-specific features in other assemblers which the GNU authors have chosen not to emulate.

One specific example (a cause of frequent questions) in the case of ARM is caused by the processor's lack of an opcode for "load 32-bit immediate into a register." There is an excellent technical reason for this: the ARM instruction word, in non-Thumb mode, is 32 bits long, and that simply doesn't leave room for an opcode and a 32-bit operand. However, the ARM assembler tries to hide this from you by directly assembling code like this without warnings:

```
ldr r1,=0x12345678
```

Of course, the ARM assembler doesn't make up a nonexistent opcode for this; it simply emulates the desired functionality by declaring a local constant. In contrast, some versions of gas will generate an error for the above code; you will have to replace it with something like this:

```
ldr r1,myconstant

{ ... more code ... }

myconstant:  .word 0x12345678
```

In addition to specialized porting-related issues like this, in order to use gas effectively you will need at least a brief introduction to the assembler's syntax.

Comments

Gas supports two classes of comment; C-style comments delimited by /* and */, and single-line comments. The character that indicates a single-line comment is target-dependent; some of the currently recognized comment characters are:

@   ARM

;   AMD 29K, ARC, PA-RISC, picoJava, PowerPC, M880x0

!   Hitachi (H8 and SuperH), SPARC, Z8000

| Motorola 680x0

Intel x86, i960, VAX, V850

Symbols and Labels

The rules for symbols (labels are just a special case of symbols) in gas are simple, but may be slightly different from the rules you're accustomed to with other assemblers. You can define a symbol either using the syntax `symbolname=value`, or as a label using the usual syntax `label:` - where label must begin with either a letter, a period or an underscore. ($ and ? are also allowable symbol characters on some platforms, but I'd advise against using them, because of the possibility for confusion.) For example,

```
mysymbol=1234
mysymbol2=5678

label:  @ This is my label
        @ (code goes here)

label2: .word mysymbol + 10
```

There is a special predefined symbol '.' (period) which represents the current value of the location counter. (In fact, the `label:` syntax is functionally identical to saying `label=.`). You can achieve various effects by manipulating this symbol; for instance, the line `. = . + 16` leaves a 16-byte "hole" in the object file, exactly the same as the .space directive. Similarly, the line `. = 0x2000` is the same as `.org 0x2000`.

Gas also implements some special constructs for local labels, useful for things such as loop points and other labels that are of no interest outside the scope of a larger function. These labels are of the form x:, where x is any positive integer (e.g. 3:, 19:, and so on). To refer to a local symbol, use the syntax xf to refer to the next (forward) reference of label x, or xb to refer to the most recent (backward) reference of label x. For example:

```
7:      b 5f
10:     b 7f
5:      b 10b
7:      b 5b
```

is exactly equivalent to:

```
a:      b c
b:      b d
c:      b b
d:      b c
```

I tend to avoid using local labels because of the possibility of inadvertently reusing the same symbol inside the middle of a long and complicated subroutine, with consequent undesirable results (and usually quite a lot of head-scratching before the problem is solved). I find that local labels are difficult to keep in mind in any function that spans more than a couple of screens' worth of code, and so I prefer to use unique and descriptive labels such as inner_copy_loop, find_string_loop, string_end_found, and so on. This is purely a matter of personal taste; if you feel confident in the use of local labels, by all means use them.

Note that if you define the same symbol twice in a single source file, the first definition overrides all subsequent definitions. Conversely, if you make reference to a symbol not defined in the current sourcefile, gas assumes that the symbol is defined elsewhere; it leaves ld to determine the value of the symbol at link time (if it isn't, you'll get an unresolved external error).

By the way, the HP PA-RISC target version of gas has slightly odd whitespace-dependent rules about labels and symbols. If you're using this highly idiosyncratic core, refer to the info page for gas for more details.

Code Sections and Section Directives

All GNU projects are divided into sections, even if some of those sections contain nothing[17]. Different sections contain different classes of code or data. Keeping your project divided into sections also makes it simple to ensure that specific pieces of code go into the right area of memory, which is particularly important in embedded systems where different memory areas have different characteristics (ROM, flash, RAM, fast on-chip SRAM, etc.). We'll go into much more detail about this later when discussing linker scripts. For the moment, we'll just cover general information about how sections work.

---

[17] Sections have nothing directly to do with segmented memory addressing models such as those found in the Intel x86 series. A section is simply a named section of memory to which code, data (or nothing at all) is emitted by the assembler.

The gas documentation defines and describes a section succinctly as "a range of addresses with no gaps; all data in those addresses is treated the same for some particular purpose. [...] 'ld' moves blocks of bytes of your program to their run-time addresses[18]. These blocks slide to their run-time addresses as rigid units; their length does not change and neither does the order of bytes within them. Such a rigid unit is called a *section*."

The three sections that gas *always* generates are named .text (typically containing code and read-only data; often stored in read-only memory or write-protected with a memory-management unit), .data (read/write variables), and .bss (uninitialized variables—i.e., RAM that is zeroed before starting the program). You tell gas where to emit any given piece of code or data using the section directive. For example, consider the following code fragment:

```
.section .text
constant1:    .word 0x12345678

.section .data
variable1:    .word 0xabcdef01

.section .bss
variable2:    .word 0
```

When assembled, this will emit one word to each of the three major sections. The word at label constant1 will be stored in the .text section, which is frequently in ROM or other write-protected memory. The word at variable1 will be in the .data section, which is in RAM.

The word at variable2 is guaranteed to contain zero at program start, regardless of whether or not you declare some value there. That's because you can't really emit anything to the .bss section—you can only reserve space in it. Anything stored in .bss is uninitialized (apart from being zeroed) at program start.

---

[18] Actually, it would probably be more accurate to say that ld moves blocks of your program to their *load* addresses. The loader or startup code in your system is where run-time addresses are handled.

When gas is assembling code, it maintains a "location counter" within each section. This location counter starts at zero[19] within each module (i.e., within each source file that is being assembled) and is incremented as code and data are emitted to the section. You can switch to a different output section to output some data, then switch back to the section you started in and continue at the same location you were before the first section switch. At link time, the various sections are gathered together (all the .text sections are merged, all the .data sections are merged, and all the .bss sections are merged) and relocation fixups are added as necessary for items whose addresses have changed, based on rules provided in the linker script.

Let's illustrate this with a more complicated program example composed of two modules. Here's partial source code for the relevant parts of module #1:

```
@@@@@@@@@@@@@@@@@@@@@@@@@@@@@@@@@@@@@@@@@@@@@@@@@@@@@@@@@@@@@@@
@ Module #1
.section .text
.code 32
.globl mysub1

        @ This subroutine sets bss2 = 0x12345678.
mysub1:    ldr r1, text1
           ldr r2, bss1
           str r1,[r0]
           bx lr

text1:      .word 0x12345678
addr_bss1:   .word bss1

.section .data
data1:      .word 0x89abcdef
```

---

[19] Purists should note that this is not the whole story. However, pretty nearly all embedded projects have to override the default placement of text, data and bss segments, so all we really need to think about is the location counter's offset from the segment's starting point. Because of this, it's much simpler to think of the location counter always starting at 0, and add in any relocation constants later.

```
.section .bss
bss1:      .word 0

end
```

When assembled, this first module will generate the following "snippets" of memory:

.text   6 x 4 = 24 bytes

.data   4 bytes

.bss   4 bytes

We can represent this as: TTTTTTDB, where 'T', 'D' and 'B' refer to one word of data in the .text, .data and .bss sections, respectively.

Now, here's sourcecode for the second module:

```
@@@@@@@@@@@@@@@@@@@@@@@@@@@@@@@@@@@@@@@@@@@@@@@@@@@@@@@@@@@@
@ Module #2
.section .text
.code 32
.globl mysub2

      @ This subroutine sets bss2 = 0x789ABCDE.
mysub2:    ldr r1, text2
           ldr r2, bss2
           str r1,[r0]
           bx lr

text2:     .word 0x789ABCDE
addr_bss2:  .word bss2

.section .bss
bss2:      .word 0

end
```

This second module will generate the following memory snippets

.text   6 x 4 = 24 bytes

.data   0 bytes

.bss   4 bytes

We will represent this as 'ttttttdb', where 't', 'd' and 'b' again represent one word of data in the .text, .data and .bss sections respectively.

Assembling these two modules, assuming that they're called module1.s and module2.s, will yield the following object files:

```
module1.o TTTTT [...] D [...] B
module2.o ttttt [...] b
```

In each object file, the .text, .data and .bss sections start at location 0.

Now, let's further assume that our linker script links module1 before module2, and that it places the .text section at some specific memory address (say, 0x02000000), the .data section immediately after it, and .bss immediately after that. (This would be a common sort of memory layout for a program that's being loaded into RAM, either by a bootloader or by a debugger). Thus, our final layout when the program is linked will be:

```
program.elf  TTTTTTttttttDBb
```

All references to items in the .text section of module2 will be fixed up (by the linker) by adding the size of module1's text section. Likewise, all references to the .bss section of module2 will be fixed up by adding the size of module1's .bss section.

The topic of link-time memory allocation is covered much more thoroughly in the section on ld, and you should read this section in detail before experimenting with the .section directive.

In the interest of completeness, I should add that the .section directive actually has a significantly more complex syntax, the details of which depend on the target executable file format (*not* necessarily the target processor type). For ELF, the target format we're using in this book, the generalized form of the .section directive is:

.section *name*, *"flags"*, *type*, *@entsize*

flags is an optional string containing one or more of the following characters:

a   section is allocatable

w   section is writable

x   section is executable

M   section is mergeable

S   section contains zero-terminated strings

If no *flags* are specified, the default behavior depends on the section name; if the name is unrecognized, the section is created with none of the above flags set, so the linker will do nothing with the section.

The optional *type* parameter can be one of "@progbits" (section contains data), or "@nobits" (section contains no data—i.e., it only occupies space in memory).

If the section *flags* specify M (mergeable) then the *entsize* parameter must be specified. The meaning of this parameter is different depending on whether or not the S (strings) flag is specified: If the S flag is specified, then the section must contain ASCIIZ strings composed of characters that are *entsize* bytes long. If the S flag is not specified, then the section must contain fixed-size constants, each of which is *entsize* bytes long.

When assembling for ELF targets, the .section directive actually pushes the current section and location counter onto a stack; you can retrieve the former context with the .popsection directive. You can exchange the current context with the context on top of the stack using the .previous directive (in much the same way as the LAST button on a television remote will switch between the current channel and the previous channel, then back again).

Arm-elf gas also supports a different syntax of the .section directive, for Solaris compatibility; this is not directly of interest to embedded developers.

Correctly handling totally custom sections with the linker is a frequent source of severe puzzlement when new programmers begin experimenting with custom section directives. Unless there are special reasons for doing otherwise, I suggest you keep to the standard section names and avoid creating totally customized sections. Most of the extended syntax around the .section directive is intended to communicate special information to scatter-loading code in an operating system. Executable loading on embedded systems is almost always implicit (the application being executed directly from ROM) or at least extremely simple (such as an application that is copied from ROM to RAM and run from RAM for improved performance).

Pseudo-Operations

Gas supports a large number of pseudo-operations for features such as reserving memory space, selecting processor-specific options, emitting text strings into your code, and affecting how the linker operates on your program. If you are coming to gas from some other assembler, most likely the gas pseudo-ops will not be the same as the ones you're used to. Here is a summary of the most commonly used such pseudo-ops, which should suffice when converting, inspecting or writing the majority of assembly-language code. Note that this list does not contain directives mentioned in the other parts of this chapter; directives related to macros, for instance, are discussed in depth in the section headed "Macros, Assembler Loops and Synthetic Instructions."

.align *boundary, fill, max-skip*

Aligns the location counter to a $2^{boundary}$-byte boundary. If you wish, you can also specify an optional fill pattern *fill* and a maximum number of bytes *max-skip* to be skipped in the alignment process; if the alignment would require more than *max-skip* bytes of padding, no alignment is performed.

Be warned that this alignment is not absolute—it is relative to the current code or data subsection, so it might not always generate exactly the final result you expect. In particular, you should be *extremely* careful about how you align and link items such as MMU page tables, complex structures that will be fetched by DMA, and other data that has to be aligned on very coarse boundaries (MMU page tables, for instance, usually have to be aligned on boundaries as coarse as 4K or 8K). To illustrate what I mean by this, consider the following code fragment, which we will call module1.s:

```
.section .text
entry_point:
    @ Perform some initialization functions

    @ Jump to rest of program
    b main_program

    @ Memory-management unit page table
    @ Our CPU requires this to be aligned on an
    @ 8K boundary.
    .align 13
mmu: .word 0x12341234
    .word 0x56785678
    @ (more MMU data follows...)
```

Let's further assume that your linker script emits the .text segment to 0x00200000. Assembling and linking just this one module will yield what you would expect: at the start of the object file you'll have the entry_point code, followed by some amount of padding, then the MMU table at an 8192-byte boundary; in the example above, the table will appear at address 0x00202000. However, this is all assuming that you were linking this code to a starting address that is already aligned properly. If your linker script emits module1's code to a starting address of say 0x00200100, then although the MMU table will still be aligned at an 8192-byte boundary *relative to the start of the .text segment in module1*, it will physically appear at address 0x00202100, which is obviously not on a physical 8K boundary.

A second subtlety of this process is fraught with even more hidden danger for the unwary. Suppose we add a second module to our project, module2.s, containing the following tiny code snippet:

```
.section .text

    @ This is a placeholder for some future
    subroutine
dummy_subroutine:

    @ Return to caller
bx lr
```

Furthermore, suppose we link these modules with a starting .text segment address of 0x00200000, in the order module2 module1 (remember that the GNU linker links object modules in the order they appear on the command line, so this module ordering is something you would control in your makefile). Module2 contains a single four-byte branch instruction, so the physical starting address of module1 will be 0x00200004. The MMU table will therefore wind up at a physical address of 0x00202004, and again it's no longer properly aligned.

As I've just illustrated, relying on the .align directive to provide guaranteed physical alignment is risky, particularly for any alignment value larger than the processor's instruction word size. You won't get any warnings about the "dealignment" that's occurring in a case like that above, and your code will probably malfunction catastrophically. I recommend that you use this directive for coarse alignment only inside a module that has a guaranteed physical starting address. A typical example of this would be your power-on initialization code.

One final note: On most platforms supported by gas, the first alignment parameter is the actual alignment byte boundary requested (i.e. .align 8192 would align to an 8K boundary). ARM, StrongARM, and i386 (a.out format) code use the syntax described above for compatibility with other, non-GNU assemblers for these platforms.

.ascii *string(s)*

Emits one or more *strings* to the current code or data section. A string in this context is a snippet of text enclosed by double-quote marks ("). Gas recognizes the following escape codes within strings:

\b	Backspace (ASCII octal code 010)
\f	Formfeed (ASCII octal code 014)
\n	Newline (ASCII octal code 012)
\r	Carriage return (ASCII octal code 015)
\t	Tab (ASCII octal code 011)

\nnn	ASCII character *nnn*, where *nnn* is a number in octal.
\xnn	ASCII character *nn*, where *nn* is a number in hexadecimal.
\\	Backslash character
\"	Double-quote character

Strings generated using this directive are not automatically zero-terminated. If you need ASCIIZ strings, you need to achieve this manually with a \000 code at the end of each string, or use the .asciz directive instead.

.asciz *string(s)*

Exactly the same as the .ascii directive, but automatically adds a zero terminator to each string.

.balign *a b c*

This directive has the same function as the .align directive, but parameter *a* is always an actual byte value, regardless of the target platform. This directive has consistent behavior across different gas-supported platforms, but it is GNU-specific.

.byte *b1, ... bn*

Emits constant data *b1 ... bn* bytewise into the current output section. Each *bx* expression is evaluated to a single byte. For example:

```
my_structure:.byte 0x01, 0x47, 0x03, 0x03, 0x12
```

.comm *symbol, length, alignment*

Declares a common symbol named *symbol*, of size *length* bytes. This is a symbol that is allocated space only once in the entire program, no matter how many modules define it. If no module actually reserves space for this symbol, the linker will allocate *length* bytes of uninitialized space.

The *alignment* parameter is an optional ELF-only parameter that specifies the desired structure alignment in bytes.

You might want to use a common symbol to communicate amongst several modules, any of which might not be included in a particular build configuration of your program. By declaring this intermodule symbol as common, including any of those modules will reserve space for the symbol without causing a duplicate symbol error at link time.

.data *subsection*

Tells the assembler to emit all subsequent code or data to the data subsection *subsection*. The *subsection* parameter may be omitted, in which case output defaults to subsection 0.

.end

Marks the end of the assembly-language program. Gas ignores any additional text after the .end directive.

.endfunc

See the entry for .func.

.endr

See the entry for .rept.

.equ *symbol, expression*

Defines *symbol* to have value *expression*. This is the same as writing simply symbol = expression, or .set symbol, expression. It is legal to redefine a given symbol many times; at any given time, the symbol has the value that was last assigned to it.

.equiv *symbol, expression*

Identical to the .equ and .set directives, except that .equiv will report an assemble-time error if *symbol* is already defined.

.err

This directive is used to signal configuration errors in conditional assembly structures.

.fail *expression*

This directive causes gas to emit either an error (if *expression* is less than 500) or a warning message (if *expression* is 500 or greater).

.float *f1, ... fn*

Emits constant data *f1 ... fn* into the current output section. Each *fx* expression is evaluated as a floating-point number. The storage format is machine-dependent; when assembling ARM code, IEEE formatting is used.

.fill *repeat, size, value*

The .fill directive has very specialized and slightly odd behavior, and appears to be of most use when porting code from some other assembler that uses this directive. The directive causes gas to emit *repeat* copies of a *size*-byte structure, starting at the current assembly location. The *size* parameter is optional (if not specified, 1 is the default value) and may be in the range 1 to 8. Any value larger than 8 is rounded down to 8. Each emitted structure is filled with data from an 8-byte internal structure, the highest-order four bytes of which are zero and the lowest-order four bytes of which are filled with *value* in the target processor's normal byte order. For example, if we are assembling for a big-endian ARM platform, the directive `.fill 2, 6, 0x12345678` would emit the byte sequence `00 00 12 34 56 78 00 00 12 34 56 78`.

Especially on little-endian architectures like x86 I find that a clearer, though GNU-specific and non-portable syntax, is to use the .rept directive along with the absolute data directives such as .byte and .word. Using this method, the above example can be rewritten like this:

```
.rept 2
    .byte 0x00, 0x00, 0x12, 0x34, 0x56, 0x78
.endr
```

.func *name, label*

Emits function call debugging information, if the file is being assembled with debugging support. The function name *name* is what you will see in the debugger as the current function if you hit a breakpoint or otherwise halt program execution. The optional *label* parameter specifies the entry-point of the function. If you don't specify a *label*, gas will use *name* prepended with a target-specific leading character. For most targets, this leading character is the underscore '_'.

For example, you might have the following assembly-language snippet:

```
        .func myfunction

_myfunction:
        @ Code goes here
        bx lr

        .endfunc
```

Note that all assembly-language functions are defined to have a void return type (in current versions of gas).

This directive has no effect unless debugging support has been specified on the gas command line. Current versions of gas only support stabs debugging information, with the —gstabs command-line switch.

.global *symbol*

Makes *symbol* visible to the linker, so that it can be referenced by external modules. By itself, this will not generate debugging information for the symbol; use the .func directive as well if you need to debug the application. For example:

```
        .globl _myfunc
        .func myfunc, _myfunc

_myfunc:  @ Code goes here
        bx lr
```

.globl *symbol*

Synonymous with .global *symbol*; see above.

.hword *h1, ... hn*

Emits "halfwords" of constant data *h1 ... hn* into the current output section. Each *hx* expression is evaluated to a 16-bit value (regardless of the target processor's nominal word size; hence the term "halfwords" is something of an inconsistency) and is stored in the target processor's default byte order. For example:

```
my_structure:   .hword 0x0147, 0x0303, 0x1234
```

.incbin *"filename",skip,count*

Includes the specified file *filename* at the current assembly location. Gas will look for the file in the search directories specified by use of the -I command-line switch.

The optional *skip* and *count* arguments are used if you only want to include a portion of the file. A typical example of this would be when embedding graphics or sound in your program; usually, you want to be able to edit these resources directly with normal image and audio editing tools. However, you often don't want to waste ROM space including header, copyright and other extraneous information from these files, and so gas gives you a way to include only a specified byte range within the file.

If *skip* and *count* are both specified, gas will seek to position *skip* bytes from the start of the file, and will read *count* bytes into your object file.

If *count* is not specified, gas will seek to position *skip* and read to the end of the file.

When using this directive, be particularly cautious about memory alignment issues after the included file, especially on architectures that cannot fetch code and/or multibyte data values from unaligned addresses. It is a prudent engineering practice to have a .align directive immediately after every .incbin directive.

.include *"filename"*

Includes the specified file *filename* at the current point in the sourcecode; the contents of *filename* will be assembled as if it occupied the file currently being assembled. The most common use for this directive is to import include files that define symbols and macros to be used in several different program modules.

Gas will look for the specified file in the search directories specified by use of the -I command-line switch.

.int *i1, ... in*

Emits constant data *i1 ... in* into the current output section. Each *ix* item is evaluated as an integer of target-specific size (in the case of ARM, the default integer size is 32 bits).

.lcomm *symbol,length*

This directive reserves *length* bytes of storage in the .bss section of the executable, and assigns it a module-local name of *symbol*. (If you want this space to be visible to the linker, and hence to external modules, you must also declare it as global using the .global or .globl directive).

By definition, this storage space will contain all zero bytes on program entry.

.list

Controls listing output. Gas maintains an internal counter that indicates whether or not a listing is to be produced. This counter is zero by default. Turning listings on using the -a command-line option increments the counter; the .list directive also increments the counter. The complementary .nolist directive decrements the counter. Listings are generated when the counter is greater than zero.

.long *l1, ... ln*

Synonymous with .int.

.nolist

See .list.

name .req reg

This directive assigns a named alias *name* to refer to the CPU register *reg*. For example, the line `heapptr .req r7` assigns the alias "heapptr" to register r7. This is an ARM-specific directive.

.octa o1, ... on

Emits constant data *o1 ... on* into the current output section. Each *ox* expression is evaluated as a 16-byte octaword (the term was coined on machines with a 16-bit word size).

.org newposition, fill

This directive moves the location counter to *newposition*. The new location counter position can be specified as an absolute expression (e.g. 0x20000000) or an expression that evaluates to a value within the current section (e.g., mylabel+0x20). If you're using the second method, be aware that since as is a one-pass assembler, you cannot .org to a currently undefined symbol. (There is a wry note in the gas documentation to the effect that if this is a serious limitation for you, you're welcome to rewrite the assembler yourself and share the result.)

.org's behavior is much more complicated than the simple ORG directive you might have used on small 8-bit assemblers, which simply sets the current output address to some specific memory location. In particular, note that the new location counter is not an absolute memory address; it is actually an offset from the start of the current section (not subsection). The actual final effect of the .org directive depends on where this particular module is linked (as the first or last module in the program), and whether this section's VMA is the same as its LMA (see the section on ld for more information on this).

Also note that .org cannot move the location counter backwards; it can only cause the assembler to skip bytes, not to "backfill" over a location already assembled.

The *fill* parameter, if specified, is the byte which will be used to pad space between the current and new location counter. If not specified, *fill* defaults to 0. For flash- and EPROM-based applications, it's normal to specify 0xFF as the fill pattern.

.p2align *boundary, fill, max-skip*

Aligns the location counter to a $2^{boundary}$-byte boundary. If you wish, you can also specify an optional fill pattern *fill* and a maximum number of bytes *max-skip* to be skipped in the alignment process; if the alignment would require more than *max-skip* bytes of padding, no alignment is performed.

.p2align is a GNU-specific, target-independent version of the .align directive, and you should carefully read the notes and instructions for that directive to appreciate more fully the subtleties of alignment requests in gas. The difference between .p2align and .align is that .align has different interpretations of the *boundary* parameter according to the target processor, whereas .p2align always interprets it in the "power of two" meaning described above. For ARM targets, the directives are functionally identical.

.p2alignl *boundary, fill, max-skip*

Functionally identical to .p2align, except that fill is interpreted as a four-byte longword value.

.p2alignw *boundary, fill, max-skip*

Functionally identical to .p2align, except that fill is interpreted as a two-byte word value.

.print *"string"*

Outputs *string* to standard output during the assembly process, for logging or special warning purposes.

.quad *q1, ... qn*

Emits constant data *q1 ... qn* into the current output section. Each *qx* expression is evaluated as an 8-byte quadword (like octaword, this term was coined on machines with a 16-bit word size).

.rept *count*

Repeats the following code block (terminated by an .endr directive) *count* times. This is extremely useful for generating tables of structures that are too complicated to implement with the .fill directive. For example, the following code:

```
.rept 4
.asciz  "Empty table entry"
.long   0x44321234
.endr
```

is equivalent to writing:

```
.asciz  "Empty table entry"
.long   0x44321234
.asciz  "Empty table entry"
.long   0x44321234
.asciz  "Empty table entry"
.long   0x44321234
.asciz  "Empty table entry"
.long   0x44321234
```

.short *s1, ... sn*

Emits constant data *s1 ... sn* into the current output section. Each *sx* expression is evaluated as a 16-bit word and stored in the target processor's default byte order.

.word *w1, ... wn*

Emits constant data *w1 ... wn* into the current output section. Each *wx* expression is evaluated as a word of target-specific size (32 bits, in the case of ARM) and stored in the target processor's default byte order.

Because gas is the usual back-end for high-level language compilers, it has some very special handling for the .word directive (in order to work around some issues that arise when generating jump tables). For this reason, I suggest that you avoid declarations of the form .word symbol1-symbol2—if you use this construct, then you risk triggering gas's special jump-table code, which will yield very unexpected results if you're assembling anything other than a jump table.

Conditional Assembly Directives

Gas supports a fairly rich set of conditional assembly directives. These operate in exactly the way you'd expect, by dividing your code into blocks that will or will not be assembled depending on the state of certain build-time conditions. Each block is opened with a conditional assembly directive that either allows assembly of the block, or causes the assembler to skip to the end-of-block marker. This marker can be one of .endif, .else or .elseif. Based on which of these end-of-block markers you use, there are three general forms of conditional assembly block. These are illustrated below:

Form 1:

```
.if condition
     (code that will assemble if condition is
     true)
.endif
```

Form 2:

```
.if condition
     (code that will assemble if condition is
     true)
.else
     (code that will assemble is condition is
     false)
.endif
```

Form 3:

```
.if condition1
     (code that will assemble if condition1 is
     true)
.elseif condition2
     (code that will assemble if condition1 is
     false and condition2 is true)
.endif
```

(Of course, in Form 3 above, you can have as many separate `.elseif` `condition`n stanzas as you want).

There are a variety of possible forms that the .if condition line can take (depending on what criteria you need to use in order to allow or disallow assembly of the code block in question), and these are described below:

.if *expression*

The following code will be assembled if *expression* evaluates to a nonzero value.

.ifdef *symbol*

The following code will be assembled if *symbol* is defined. This form of conditional assembly is most commonly used in conjunction with the --defsym command-line switch when using different makefiles to build specialized versions of a particular piece of code. For example, consider the following code:

```
.ifdef DEBUGGING
    @ Output debugging messages
.endif
```

The debugging messages will be assembled into the program if gas is given the command-line option `--defsym DEBUGGING=1`.

.ifc *string1,string2*

The following code will be assembled if *string1* and *string2* are identical (this is a case-sensitive comparison). Single quotation marks around the strings are optional; if no quotation marks are used, then *string1* ends at the first comma, and *string2* ends at the end of the line.

.ifeq *expression*

The following code will be assembled if *expression* evaluates to zero.

.ifeqs *string1,string2*

Similar to .ifc, except that *string1* and *string2* must be enclosed in double quote marks.

.ifge *expression*

The following code will be assembled if *expression* is greater than or equal to zero.

.ifgt *expression*

The following code will be assembled if *expression* is greater than zero.

.ifle *expression*

The following code will be assembled if *expression* is less than or equal to zero.

.iflt *expression*

The following code will be assembled if *expression* is less than zero.

.ifnc *string1,string2*

Follows the same rules as .ifc, but the result of the string comparison is inverted; the following code will be assembled only if *string1* and *string2* do not match.

.ifndef *symbol*

The following code will be assembled if *symbol* is not defined.

.ifnotdef *symbol*

Synonymous with .ifndef.

.ifne *expression*

The following code will be assembled if the expression evaluates to value other than zero. This is functionally identical to the .if directive.

.ifnes *string1,string2*

.ifnes is to .ifeqs what .ifnc is to .ifc; it will assemble the following code only if string1 and string2 do not match.

Macros, Assembler Loops and Synthetic Instructions

In addition to the directives explained above, gas supports a macro system in order to reduce your typing and debugging workload. Macros are declared using the .macro directive, and end with the .endm directive. The general form of a macro definition is as follows:

> **.macro** *macro-name macro-arguments*
>   (code and/or assembler directives)
> **.endm**

(Although all macros must end with a .endm directive, you can bale out of any macro early using the .exitm directive; note that this only exits the innermost layer of a recursive macro like the one illustrated below.)

The second argument is optional; if your macro needs to take parameters, you should specify their names here, separated by commas. If a parameter needs to have a default value when the caller supplies none, you can achieve this by appending *=value* to the desired parameter name. For example:

```
.macro mymac param1=0,param2,param3
```

specifies a macro named "mymac", with three parameters. If the caller supplies no value for param1, a default value of 0 will be used within the macro.

Within the macro, you refer to parameters with the code *parameter-name*. For example, here is a simple macro, taken directly from the gas documentation, that emits an increasing series of 32-bit numbers (within specified boundaries):

```
.macro storenums from=0, to=5

@ Store the current value of from at the
@ current location
.long \from
```

```
@ If to != from, then call down recursively
.if \to-\from
    storenums "(\from+1)",\to
.endif

.endm
```

You invoke a macro simply with its name, followed by its arguments if applicable. For instance, invoking the macro above:

```
@ Invoke storenums macro with default settings
storenums
```

will yield the following code:

```
.long 0
.long 1
.long 2
.long 3
.long 4
.long 5
```

(Because we didn't specify any parameters, default values were used).

When invoking a macro, arguments can be specified by position (using a comma-delimited list in the same order as the parameters in the macro definition) or by name. For example, given the macro definition above, the following two invocations are identical:

```
storenums 1,6
```

and

```
storenums to=6, from=1
```

You can "undefine" a macro using the .purgem directive:

.purgem *macro-name*

Any reference to *macro-name* after the .purgem directive will not be expanded.

As well as the user-defined macro system described above, there are also a couple of directives that are, to all intents and purposes, predefined macros. These are detailed below:

.irp *symbol, value-list*

This directive assembles a block of code one or more times, assigning a different value from *value-list* to *symbol* on each pass. The code block must be terminated by a .endr directive. Inside the code block, you can retrieve the value of symbol for the current assembly pass using the syntax *symbol*. Values within the list are comma-separated.

For example, the following code will load 0x00000000 into registers r0 through r6:

```
.irp regnum,0,1,2,3,4,5,6
    ldr r\regnum, =0
.endr
```

It decomposes to the following:

```
ldr r0, =0
ldr r1, =0
ldr r2, =0
ldr r3, =0
ldr r4, =0
ldr r5, =0
ldr r6, =0
```

This pseudo-instruction is designed to simplify tasks such as fetching several function arguments from the stack. It is not a vital feature on cores such as ARM, which have hardware instructions to load multiple registers automatically with auto-increment/decrement of a source pointer register, but it can be useful.

Note that if you don't specify any values, the .irp-encapsulated code block will be assembled once, with *symbol* evaluating to an empty string. This will usually cause assembly errors.

.irpc *symbol, character-list*

This is a special cut-down version of the .irp directive. It works the same way as .irp, except that on each pass, the symbol is assigned a single character value from *character-list*. (The .irp directive, in contrast, assigns an arbitrary-length string value to *symbol*). *Character-list* is a simple string of characters, not separated by commas or other punctuation.

The example above could be rewritten using the .irpc directive thus:

```
.irpc regnum,0123456
    ldr r\regnum, =0
.endr
```

Again, if you don't specify anything in *character-list*, the code block will be assembled once, with *\symbol* evaluating to an empty string.

Ld—GNU Linker

Introduction

Ld is the GNU linker. It is an immensely powerful tool which allows you to control the positioning and attributes of object code in the final output file, and the virtual memory map of the result (these are not always the same thing, as we will see later). You can control these details using a plain-text file called a linker script, typically given an extension of .ld. Because of the sheer flexibility of this tool, and the numerous interacting options, it is extremely difficult to choose a logical order in which to describe the linker's features. Ld has an astonishingly simple syntax, but an equally astonishingly large number of interacting, equivalent or complementary syntactic constructions. Some of these are due to the unavoidable fact that it is rather an old program with a corresponding degree of design cruft, but mostly these constructs are present either to ease porting code from some other developme toolchain, or in order to support special executable file generation needs in various operating systems. The same basic linker constructs are expected to be able to generate Win32 Portable Executable files, ELF executables, COFF executables, old a.out executables and raw binary ROM dumps for embedded systems (among many other formats), so you can see why the syntax is complex.

Thus, I strongly suggest you read this section in its entirety to obtain a reasonably good overview of how to control ld. Frankly, it is rather difficult to discuss this tool unless you have specific goals in mind, which is why it will probably be most useful to you to pick a linker script that does something similar to what you need already, and modify it to suit your requirements. You can then refer to this chapter and the on-line documentation for help when you encounter a problem. When I was initially learning how to use the GNU toolchain, almost all of my learning curve was spent learning how to use ld effectively, and I found that by far the easiest way to get familiar with the concepts (after initially reading up on the basic script syntax) was to look at example scripts along with simple projects, change the scripts, relink the program and inspect the resulting changes using objdump.

In order to understand why it's important for us to have such an awesomely flexible linker, let us first consider the types of information that are typically combined into a single bootable ROM image:

- **Startup** (hardware and C run-time initialization) code. This code is practically always written in assembly language and must be located at a specific place in ROM.

- **Application code.** This is quite distinct from startup code and usually doesn't have to reside at any specific area in the memory map.

- **Constant data** (e.g., a C constant declared as `const char mystring[] = "My string"`, binary files included in your ROM image, and so on) This type of information can safely be stored in ROM and used in situ.

- **Initialized variables** (e.g., a C global variable declared as `int myint = 1234`). Although this data must physically reside in RAM, the initial values to be loaded at boot time must be in ROM.

- **Uninitialized variables** (e.g., a C global variable declared simply as "int myint"). These don't need to occupy any space in ROM at all; the startup code simply needs to allocate sufficient RAM space for them, and the linker needs to know how to resolve references to these variables.

The situation is further complicated by the fact that in embedded systems, the memory map usually changes drastically during execution of the startup code; this power-on initialization code often remaps chip select lines, enables a memory-management unit or otherwise changes the system layout.

For this reason, one of the first important concepts you should understand is the difference between "load memory address" (LMA) and "virtual memory address" (VMA). The VMA acronym has nothing directly to do with the idea of virtual (vs. physical) memory and complex demand-paging virtual memory systems. It simply refers to the difference between the location of a piece of code or data in the executable file (LMA) vs. the location in memory where references to this code or data should be directed (VMA).

For example, consider the .data section of a piece of code that runs out of ROM. Before running the main program, the bootstrap code needs to initialize the .data section by copying information out of ROM into the RAM space intended to hold it. Obviously, though, all coded references to symbols in the .data section need to refer to the RAM copies, because the ROM copies are just preload data and can't be modified. In ld parlance, the .data section is linked with the LMA in ROM and the VMA pointing to the real RAM versions.

There are basically two ways of informing ld where to load the various parts of your program. The first is to assign names to the various regions of memory in your device, and then direct each code or data section to the appropriate memory region. The second method, which is less "friendly" but potentially slightly less ambiguous, is to start the linker's current memory location counter at a known address (the start address of the first section of memory to be populated) and emit sections one by one to the current location, manually incrementing this location counter as appropriate in order to skip unpopulated "holes" in the memory map.

Let's discuss the latter method first, since it is conceptually slightly easier to digest. Here is a very simple linker script that you might use for a program for the Atmel EB40.

```
/* Example minimalist linker script for
   Atmel EB40 */
SECTIONS
{
    . = 0x02000000;
    .text : { *(.text) }
    .data : { *(.data) }
    .bss : { *(.bss) }
}
```

Let's deconstruct this linker script in order to get a basic introduction to ld's commands.

First, observe that C-style comments delimited by /* and */ can be included in linker scripts to improve comprehensibility.

Next we have the SECTIONS command. SECTIONS (followed by a list of output sections enclosed in curly braces) is the command that tells ld how to output the destination file. The first line in this stanza sets the value of the location counter. Just as in gas, the location counter is a special variable with the name '.', and you can update it or calculate with it if you need to. By default, the location counter starts at 0. It is incremented (and aligned, if necessary) each time you send some output to the destination file. The first thing we need to do, therefore, is set the location counter to point to the start of the EB40's RAM, as that is where we intend to load our program. The line `. = 0x02000000;` achieves this.

Having done this much, we need to tell ld which sections to include in the output file, where to emit them into memory, and which sections of the input file(s) should be mapped. The next three lines of the script perform this task. Basically, these lines say "collect all .text sections from the input files and emit them to a section called .text in the output file. Then collect all .data sections from the input files and emit them to a section called .data in the output file. Finally, collect all .bss sections from the input files and emit them to a section called .bss in the output file".

Let's go into more detail for each of the commands and subcommands used in the above example.

The SECTIONS command

The SECTIONS command has a much more general application than the above example. The general format is:

```
SECTIONS
{
        sections-command #1
        sections-command #2
        . . .
        sections-command #n
}
```

Each sections-command line can be an ENTRY command, a symbol assignment, an output section description or an overlay description. Symbol assignments, output sections and overlay descriptions are described in their own separate sections below.

The ENTRY command is used to inform whatever loads your program—gdb, in our case—what the entry point for your program is. Ld sets the entry point to be one of the following (in decreasing priority order):

- A value supplied on the command-line with the "-e SYMBOL" switch.

- The value set with an ENTRY(SYMBOL) command in the linker script.

- The value of the symbol "start", if defined.

- The address of the first byte of the .text section, if there is one, or

- Zero.

So, if we want our program to have the entry-point "vectors", we would add the line:

```
ENTRY(vectors)
```

to the linker script. However, the ENTRY command doesn't have to reside within a SECTIONS stanza (the examples given in this book, in fact, have it in the main script body towards the start of the file). The reason ld allows you to put this command inside the SECTION definitions is so that you can assign the entry-point a value based on some calculation with an intermediate value of the location counter, such as ENTRY(. + 0x100).

As a matter of largely academic interest, you can write a very minimalist linker script containing no SECTIONS stanza at all. In this case, the output file will have a starting address of 0, and the segments in it will be collected, and ordered in the same way they are first encountered. For instance, if the first object module linked contains all of your program's sections, the output file will have its sections ordered the same way as that module. If the first module has, say, only .text and .data sections (in that order), and some subsequent module has a .bss section, then the output module will have .text and .data sections followed by a .bss section. The output format is further affected by the attributes of any named memory regions you have defined in the linker script (see "Named Memory Regions" below). As you can imagine, this ill-specified output format is not terribly useful, at least in embedded applications that lack a complex loader/relocator module.

Symbol Assignments, Expressions and Functions

Symbol assignments are of the form *symbolname = value*; (the trailing semicolon is mandatory). Any symbol you define in the linker script is automatically global. You can refer to these symbols in your program, and indeed it is very useful to be able to do so. For example, it is common to see a SECTIONS statement something like this:

```
SECTIONS
{
        . = 0x02000000;
        .text : { *(.text) }
        .data : { *(.data) }
        _bss_start = . ;
        .bss : { *(.bss) }
        _bss_end = . ;
}
```

In your power-on initialization code, when you want to zero out the contents of the .bss section, you could have code like this:

```
.globl _bss
.globl _ebss

@ Clear .bss section
ldr r1, bss_start
ldr r2, bss_end
ldr r3, =0
clrbss:
    cmp r1,r2
    strne r3,[r1],#+4
    bne clrbss

{ ... other initialization code ... }

    @ Jump to main C program
    bl main

    @ Hang system if main program returns
    b .

bss_start:
    .word _bss

bss_endptr:
    .word _ebss
```

This way, as you change your program and the size of the .bss section shrinks and grows, your startup code doesn't need to rely on any hardcoded constants—it's automatically kept up to date by the linker.

Note that you can do a lot more than simply give a symbol a numerical value; ld supports a rich variety of arithmetic operators and C-style assignments. The following assignment operators are available (these work exactly like their C counterparts):

Assignment:	expression1 = expression2 ;
Addition:	expression1 += expression2 ;
Subtraction:	expression1 -= expression2 ;
Multiplication:	expression1 *= expression2 ;
Division:	expression1 /= expression2 ;

Left shift:	expression1 <<= expression2 ;
Right shift:	expression1 >>= expression2 ;
Logical-AND	expression1 &= expression2 ;
Inclusive-OR	expression1 |= expression2 ;

In addition to these assignments, you can use the equivalent arithmetic operators +, -, *, /, <<, >>, & and | in arithmetic expressions. Ld evaluates expressions following C syntax rules, and as 32-bit values (on 32-bit platforms; 64-bit values on 64-bit platforms).

As well as the above, ld features an alternate way to define symbols using the PROVIDE command. The need for and usage of this keyword are somewhat esoteric but you might encounter it in other peoples' linker scripts, so you need to understand what it does. The general syntax for this command is PROVIDE (*symbol = expression*); . *Symbol* will be set to the value of expression only if *symbol* is referenced in your program **but not defined in any module**. So for instance, consider that you have a script fragment like this:

```
SECTIONS
{
    .data :
    {
      *(.data)
      _edata = .;
      PROVIDE(edata = .);
    }
}
```

If your program defines a symbol called _edata, the link operation will fail with a duplicate symbol error. If your program defines a symbol called edata, the linker will use the definition in your program. If your program references but does not define the symbol edata, the linker will use its internally generated definition.

The symbol '.' is always defined; as discussed earlier, it contains the current value of the location counter. You can perform any kind of arithmetic with this, just as with any other symbol; you can assign the location counter a new value, or use the current value to calculate some other result. The only restriction, as in gas, is that you can only move the location counter forwards. If you need to overwrite the same memory area with different sets of data for some reason, you should look at ld's syntax for handling overlay sections, described below.

Very important note: The location counter retrieved when you reference the '.' symbol is relative to the start address of the current enclosing object; **it is not an absolute address**. For example, look at the following script fragment:

```
SECTIONS
{
        . = 0x1000;
        .text :
        {
                *(.text)
                . = 0x2000;
        }
        .data :
        {
                *(.data)
        }
}
```

This fragment will generate an output file containing the .text section at a starting address of 0x1000. This section will be padded to a size of 0x2000 bytes (*not* 0x1000 bytes, which is what you would expect if '.' referred to an absolute location counter). The .data section starts at address 0x3000.

This is actually a specific case of a more general rule—all expressions in ld generate either relative or absolute results. Expressions within an output section definition generate relative results (relative, that is, to the start of the output section). Relative results have a "sectionness"; they are associated with their parent section. Expressions that are not within an output section definition generate absolute results that have no dependency on any section. Relative symbols can be made relocatable by specifying the -r option on the ld command line; absolute symbols can never be relocated.

If you need to make a specific symbol absolute, you can use the ABSOLUTE keyword. For example, the line `mysymbol = ABSOLUTE (.);` will assign the current absolute value of the location counter to mysymbol, even if this assignment appears within an output section description.

ABSOLUTE is one of ld's assorted built-in functions. The remainder of these functions are listed below (less a couple of functions that have limited relevance to embedded applications):

ADDR(*section-name*)

Returns the absolute VMA of the starting address of section *section-name*. This function can only return the address of a section that has previously been defined; it cannot resolve forward references.

ALIGN(*expression*)

Returns the value of the location counter aligned to the byte boundary specified by *expression*. Note that *expression* must be a power of two. In order to actually use this result, you will most often see ALIGN used in a statement like this:

```
. = ALIGN(4);
```

BLOCK(*expression*)

This is a synonym for ALIGN(*expression*).

DEFINED(*symbol*)

Returns 1 if *symbol* is a defined global, or 0 otherwise. The canonical example where you might want to use this function is to define a symbol only if it is not already defined, as in the following example:

```
SECTIONS
{
    .text :
    {
        entry_point = DEFINED(entry_point) ?
          entry_point : . ;
    }
}
```

This assigns the value of the location counter to the entry_point symbol, as long as this symbol was not already defined. If the symbol was already defined, it is assigned to itself, meaning that it is effectively left unchanged. (Note the C-style syntax there, by the way).

LOADADDR(*section-name*)

Returns the absolute LMA of the starting address of section *section-name*.

MAX(*expression1, expression2*)

Returns the larger of *expression1* or *expression2*.

MIN(*expression1, expression2*)

Returns the smaller of *expression1* or *expression2*.

NEXT(*expression*)

Returns the next unallocated memory address that is an integral multiple of *expression*. The only difference between NEXT and ALIGN is that if you have defined named memory regions using the MEMORY command, ALIGN will simply perform some address arithmetic (which may yield a result outside the boundaries of any memory region), whereas NEXT will skip to the next defined region, if necessary.

SIZEOF(*section-name*)

Returns the size, in bytes, of section *section-name*.

Output Section Descriptions

We have been making frequent use of output section descriptions above without really defining their syntax or how they work, and now it's time for some more detail. The general format for an output section description is:

```
output-section address (type) : AT (lma)
{
     output-section-command #1
     output-section-command #2
     ...
     output-section-command #n
} >region AT>lma-region :phdr =fillexp
```

output-section is the name of the desired output section within the output file; for example, .text. The optional *address* parameter is the starting value for the location counter (VMA) in this section. If no value is specified, the last value of the location counter is used, with some padding bytes for alignment if the output section type requires a specific alignment value. (The default value at the start of the script is 0, as you recall from above). If you provide a specific value for *address*, this overrides the section's alignment requirements. Note that *address* can be any arbitrary expression; for example, .text ALIGN(0x40) : { *(.text) } is a perfectly valid output section description.

Similarly, the optional *lma* parameter is the desired starting LMA for the section. If this parameter is not specified, the LMA is assumed to be the same as the VMA.

The *type* parameter, which again is optional, specifies the output section's characteristics. In current versions of ld, there are only two different types:

NOLOAD

Specifies that the section should be marked as not loadable, so that it will not be loaded into memory when the program is executed.

DSECT, COPY, INFO, OVERLAY

All of these keywords are synonyms for backward compatibility. They all specify that the section should not have any space allocated for it at load time.

You probably won't use the *type* parameter in an embedded program.

Moving on, the first two parameters (*region* and *lma-region*) after the closing curly bracket are equivalent to the *address* and *lma* parameters, except that they refer to named memory regions rather than specific addresses. If you specify an output region and/or LMA region at the end of the stanza, you do not need to specify output addresses or LMA at the beginning of the stanza, and vice versa. We'll discuss regions in more detail a little later (refer to the section headed "Named Memory Regions" below). For the moment, just note that the *address* and *lma* parameters take precedence over any region output instructions specified with the *region* and *lma-region* parameters.

The *phdr* parameter is used to manipulate ELF program header information; it assigns a section to a program segment. This option is an example of ELF esoterica that you are unlikely to need to explore in an embedded system.

Finally, the *fillexp* parameter specifies the value to use when filling memory areas not explicitly occupied by code or data. If you specify *fillexp* as a simple hexadecimal number (the letters 0x followed by valid hexadecimal digits) then you can specify a fill pattern of arbitrary length. If you enclose the *fillexp* parameter in parentheses or add a unary plus sign, the parameter will be interpreted as a 32-bit constant. Regardless of how you specify *fillexp*, the fill pattern is stored in big-endian format (in other words, the same order as you specified it).

Note that if our input files contain some sections for which we haven't specified any particular handling—for example, symbol table debugging information in a .stabs section—these would be included in the output executable with default names and output addresses. If you explicitly want to discard data from the input files at link time, you can achieve this by including an output section directive that places all the unwanted data in a special section named /DISCARD/—for example, the following instructs ld to throw away debugging information during the link process:

```
/DISCARD/ { * (.stab .stabstr) }
```

Overlay Section Descriptions

An overlay section is used when you want to have several items of code or data located on top of each other in the same memory area. The assumption is that you have some external overlay manager code that will copy the required data section from its LMA to the area of RAM reserved for overlays as necessary.

Overlays have a couple of uses in embedded systems. One example is if you have a rather large program and you have neither ROM nor RAM space to store it all in uncompressed form. Another example is if your program is stored in a slow, cheap type of memory and you want to execute it out of faster RAM. As long as you can modularize your code sufficiently, you can keep most of it compressed in ROM and only uncompress to RAM the specific module you need at the moment.

An example of either of these scenarios might be a simple PDA type of project containing, say, a word processor and a spreadsheet application. The application code might be stored in a slow memory type, such as NAND flash memory. When the user wants to edit a memo, your overlay manager copies the word processor code from flash to RAM and runs it. If the user then wants to edit a spreadsheet, the overlay manager can tell the word processor code to save the current state of its document, then it can kill the word processor task, copy the spreadsheet program from flash memory to RAM (overwriting the word processor code) and run the spreadsheet.

Overlays are specified using the OVERLAY command. This command is used in the same context as an output section directive; i.e., within a SECTIONS command. To put it in C terms, you can think of an output section directive as a struct (every member occupies its own space) and an OVERLAY command as a union (several different structures are defined to occupy the same space). The general format of the OVERLAY command is as follows:

```
OVERLAY start-address : NOCROSSREFS AT (lma)
{
     section_name_1
     {
          output section command #1
          [...]
          output section command #n
     } :phdr =fillexp
     [ ... arbitrary number of sections ... ]
     section_name_n
     {
          output section command #1
          [...]
          output section command #n
     } :phdr =fillexp
} >region :phdr =fillexp
```

The optional *phdr* and *fillexp* parameters, if specified, have the same meaning as they do for output section directives (see above).

Each section declared inside the overlay definition is given the same starting VMA. The sections are emitted to increasing LMAs, one after the other. You can specify the starting LMA for the overlay section as a whole using the optional AT (*lma*) parameter; if you do not provide this parameter, ld will assume that you want the LMA to be the same as the VMA of the start of the structure. You can specify the starting VMA for the structure with either the optional *start-address* parameter, or by specifying a named output memory region with the optional *region* parameter. In exactly the same way as for normal output sections, if you don't explicitly provide a *start-address* or output *region*, ld will by default use the current location counter. You *cannot* specify LMAs or VMAs for the individual sections within the overlay description, nor direct them to named memory regions.

If you add the optional *NOCROSSREFS* keyword, ld will check for references between the code and data included in each section within the overlay definition. Any cross-references of this type *almost* always indicate a programming error, because by definition, all the sections in an overlay are loaded to and executed from the same memory space. There are, however, scenarios when you might conceivably want to refer to something that's in a swapped-out overlay "underneath" your code. For example, you might want to generate a pointer to some piece of code inside a different overlay, so that you can tell the overlay manager "swap me out, swap in overlay #2, and jump to this specific address". Being able to suppress cross-referencing errors allows you to do this without the manual tedium of having to create external pointer tables in memory that is never swapped out.

When you define an overlay section, ld will automatically generate some symbols that will be needed by your overlay manager. These symbols are of the form __load_start_*sectionname* and __load_stop_*sectionname*, and they contain the starting and ending address, respectively, of the LMA of the named section within the overlay description. This can be illustrated best by use of an example; let's consider the PDA application mentioned above. We'll assume that this PDA has a small amount of fast RAM at address 0x00100000, and that we have 128K of very slow flash memory or ROM at address 0x00000000. (It's not unknown for very low-end microcontroller type applications to use excruciatingly slow *serial* ROM for cost reasons, so this is a fairly realistic project.) Let's further assume that we are going to reserve the first 64K of ROM for the bootstrap code, overlay manager and other operating system intricacies, and use the second 64K for our overlays. We could define the overlay with something like this (assuming that the object files for our word processor are all in the wordpro/ directory, and the object files for our spreadsheet are in the spreadsheet/ directory):

```
_OVERLAY_AREA = 0x00100000;

OVERLAY 0x00100000 : AT (0x00010000)
{
        .text0 { wordpro/*.o (.text) }
        .text1 { spreadsheet/*.o (.text) }
}
```

The word processor code and constant data will be stored at address 0x00010000 in ROM, and the code for the spreadsheet will be emitted immediately after it. Both pieces of code, however, expect to be relocated to 0x01000000 before they are executed. In order to run the word processor, we could write a subroutine in ARM code something like this:

```
_run_wordprocessor:
        @ Copy word processor code overlay to RAM
        ldr r0,overlaystart
        ldr r1,overlaydest
        ldr r2,overlayend

copy_overlay:
        cmp r0,r2
        ldrne r3, [r0], #+4
        strne r3, [r1], #+4
        bne copy_overlay

@ Call the code just copied
        bl _OVERLAY_AREA

@ Return to our caller
        bx lr

overlaystart:
        .word __load_start_text0
overlayend:
        .word __load_end_text0
overlaydest:
        .word _OVERLAY_AREA
```

(Note that you might need to do a little more work than this for a real application, since you probably want to swap out the data sections for the program as well.)

By the way, there's no technical reason why you would have to do this in assembly language; you could do it just as easily in, say, C by defining a global function pointer to the start of the overlay region, referencing the __load_start_* and __load_end_* symbols as external pointers to type const char (e.g. `extern const char * __load_start_text0;`) and using memcpy to copy the overlay from ROM to RAM.

Emitting Data Directly into the Executable

Within an output section description (only!), you can directly emit constant data to the output file using the following five commands:

BYTE (*expression*)

Emit a byte (the result of evaluating *expression*) to the output file.

SHORT (*expression*)

Emit a 16-bit word (the result of evaluating *expression*) to the output file.

LONG (*expression*)

Emit a 32-bit longword (the result of evaluating *expression*) to the output file.

QUAD (expression)

On 64-bit platforms, emit a 64-bit quadword (the result of evaluating *expression*) to the output file. On 32-bit platforms, emit a 64-bit quadword, the result of evaluating *expression* as a 32-bit value and zero-padding it to 64 bits.

SQUAD (*expression*)

On 64-bit platforms, emit a 64-bit quadword (the result of evaluating *expression*) to the output file; this is identical to the behavior of QUAD (*expression*). On 32-bit platforms, emit a 64-bit quadword, the result of evaluating *expression* as a 32-bit value and sign-extending it to 64 bits.

FILL (*expression*)

This command sets the fill pattern for unused memory areas. The difference between FILL(xxxx) and using the =*fillexp* parameter on the output section description is that the =*fillexp* parameter applies to the entire section from start to end, whereas the FILL command takes effect only from the point in the file where you specify the command.

Note: If the output file format has a specific endianness (e.g., ELF), the data emitted by the above commands will be stored with that endianness. A few supported output file formats, such as Motorola S-records, do not have an explicit endianness; in which case, ld will store data declared this way with the same endianness as the first object file linked into the executable.

These directives are useful for assembling special structures such as jump tables, interrupt vector tables, and so on. With appropriate use of these directives, you can elegantly automate the generation of all such tables, so that as your code and data change in size, you never have to manually calculate the contents of these tables.

Input Section Descriptions

An input section description is a filename (e.g. module1.o) optionally followed by a list of section names in parentheses. Both file name and section name may be wildcarded if desired. The examples we used above, without explanation, are far and away the most common usage of the input section command. The general form we were using is:

> * (*input-section-name*)

where *input-section-name* is the same as the name of the surrounding output section description, and the * wildcard means "process all input files". For example, the command * (.text) means "now emit the .text sections of all input files".

This is all you need for the vast majority of cases. However, there are specific instances where you'll want to output code from a particular module into a different area. For instance, you might want to ensure that power-on startup code goes at the very start or very end of a ROM image (according to your processor's requirements). If this module is called boot.o, you might have a linker script something like this:

```
SECTIONS
{
      .init : { boot.o (.text) }
      .text : { *(EXCLUDE_FILE (*boot.o) .text) }
      .data : { *(.data) }
      .bss : { *(.bss) }
}
```

Note the EXCLUDE_FILE syntax. This prevents ld from trying to include the boot.o code in the main .text section.

Ld supports more complex wildcards than simply the catch-all asterisk; it supports a subset of normal UNIX shell wildcards:

*	matches anything
?	matches any single character
[ABCD]	matches any character specified within the square brackets.
	A range of characters can be specified with the - character, e.g. [A-Z] matches any alphabetic character.

Wildcards will not match the UNIX path separator, '/'. Also note that ld does not search directories for wildcarded filenames; it only searches, and will only match, filenames that it has been "told about" on the command line or using the INPUT directive.

Alternatively, you might want to collect two input sections together for some reason. For instance, suppose you wanted to emit both the .data and .bss sections of the input files to the .data section of the output file. (There doesn't appear to be an obvious application for this, but it's perfectly legal. The easiest way to work with such a configuration would probably be to zero the entire RAM area and then copy the initialized data across from its LMA to its VMA in the data segment.)

Whatever your reasons, you could specify this either as:

```
.data : { *(.data) *(.bss) }
```

or

```
.data : { *(.data .bss) }
```

The difference between these two definitions is that in the first case, ld will collect all the .data sections from the input files together and emit them to the output file, followed by all the .bss sections. In the second case, ld will merge items from the .data and .bss sections in the order it encounters them in the input files.

Named Memory Regions

One downside to the above technique for writing linker scripts is that ld knows absolutely nothing about the system's actual memory map and hence you get no feedback if you accidentally over-reach the physical boundary of some memory area. For example, if you add a large uninitialized data structure to your program, the end of the .bss section might now protrude beyond the end of physical RAM. When the power-up code attempts to zero this memory, because of incomplete address decoding, the end of the write operation will wrap around to the bottom of RAM and destroy part or all of the program.

You can avoid this problem, and get ld to work a little harder for you, by defining the memory map of your target using the MEMORY command to assign names to specific regions. For instance, we can define the relevant section of the EB40's memory map using this stanza:

```
MEMORY
{
    sram : org = 0x02000000, len = 0x00080000
}
```

This assigns the name "sram" to the memory range from 0x02000000 to 0x0207FFFF, which corresponds to the 512K of off-chip SRAM on the EB40. If we include the above definition, our simple minimalist linker script for the EB40 becomes:

```
SECTIONS
{
    .text : { *.(text) } >sram
    .data : { *.(data) } >sram
    .bss : { *.(bss) } >sram
}
```

This is the syntax used in the linker scripts for the examples in the next chapter. The generalized format of the MEMORY directive is:

```
MEMORY
{
    name attributes : ORIGIN = origin, LENGTH
        = len
    [ ... an arbitrary number of additional
        named regions may follow ... ]
}
```

The *name* parameter is an arbitrary, user-chosen friendly name that describes the region, such as "rom", "static_ram", "dram_bank_1" and so on. *Origin* is the starting address of the memory region; this must evaluate to a constant before any memory allocation is performed, so it cannot use any symbols that depend on section lengths. Likewise, *length* is the size of the memory region, in bytes. This can be specified using standard abbreviations such as "64K" or "8M". (Note also that the keyword ORIGIN can be abbreviated to org or o, and the keyword LENGTH can be abbreviated to len or l.)

The optional *attributes* parameter is only used if you do not specify a SECTIONS directive in your linker script. It consists of a string of attribute characters with the following meanings:

A	Allocatable section
I	Initialized section
L	Loadable section (synonymous with I)
R	Read-only section
W	Read/write section
X	Executable section
!	Inversion operator (e.g. "!R" means "not read-only")

If one of the input files contains a section that is not explicitly directed using an input section directive in the linker script, and the attributes of that section match those of a defined memory region, ld will direct the unmapped section to that memory region.

Special Considerations for C++

If you're writing in C++, you should be aware that GNU C++ generates some extra sections and requires special handling for global constructors and destructors. Gcc places a table of constructor call pointers into the .ctors and .dtors sections in your input object files, respectively. You need to collect this information into a structure in your executable so that the run-time initialization code can run all the constructors at program start-up and all the destructors when the program exits. This structure is slightly more complicated than just a raw list; the first word in the table is the number of entries, and the last word is zero. In order to emit this table automatically into the executable, you need a construct like that below in your .text output section:

```
__CTOR_LIST__ = . ;
LONG((__CTOR_END__ - __CTOR_LIST__) / 4 - 2)
*(SORT(.ctors))
LONG(0)
__CTOR_END__ = . ;

__DTOR_LIST__ = . ;
LONG((__DTOR_END__ - __DTOR_LIST__) / 4 - 2)
*(SORT(.dtors))
LONG(0)
__DTOR_END__ = . ;
```

(By the way, the reason we subtract two from the table-length value is because the calculation above includes the table-length word and the zero terminator word.)

This book does not discuss GNU C++ programming in detail, partly because C++ adds (often unnecessary) complexity to almost any project, and partly because I personally belong to the school of thought that believes C++ is usually not a good choice for embedded systems programming; using C++ (as opposed to procedural C) requires significantly more rigor in design to achieve equally reliable and functional results as an equivalent C program. Programs that absolutely require C++ also tend to be those projects that will always be built around a pre-existing operating system. This operating will have its own rules about executable layout, and the cross-development toolchain for the operating system will certainly include all the linker control magic you need to generate correctly formatted executables. As a result, I won't go into a lot of detail on this topic.

Further ld Information

Almost every option that can be specified on the command line for ld can be coded into the linker script itself, and vice versa. This allows you to balance where your settings are coded; either into the makefile for the project, or into the linker script. My personal preference is to specify board-specific material in the linker script, as far as possible, and project-specific material in the makefile. Whatever you decide to do, it's best to pick a set of general rules and stick with them rigidly, otherwise when you come to reuse a linker script, you may find that having specialized defaults coded into it causes subtle problems with other projects.

For more information on the ld script format, you should refer to the info page for ld. The information given in this chapter is a fairly complete introduction to the syntax and capabilities, but the on-line documentation includes more illustrations and describes a lot of directives that I don't feel are absolutely necessary reading for the first-time embedded GNU developer. Basically, if you can think of a need to order specific sections of your code in a particular way, there is almost certainly a linker script syntax to achieve what you want to do—actually, there are probably at least three easy ways of doing it, and several more circumlocutive ways of achieving the same thing.

By the way—purely as a matter of interest—contrast ld's vast smorgasbord of options with the capabilities of linkers built into operating-system-specific compilers. Such linkers are hardwired to generate code with load and run addresses specific to the properties of the target operating system's memory manager and load-and-execute code. This is one reason why general-purpose compilers—or at least the linkers that ship with them—aren't easily retargetable to generate embedded code.

Converting Files with Objcopy

The function of objcopy is to manipulate binary files, usually structured files such as ELF executables. You will use it most frequently when converting from one file format (such as ELF) to another (such as a raw binary dump to be written into ROM). A pass through objcopy is usually the last stage in most embedded compilation jobs. You can also use objcopy to shuffle segment load addresses and add or remove sections if necessary, but there should rarely be a need for this. If you need to tweak the load address of a section in your program, the best place to do this is in the linker script.

The syntax you will see used most often for objcopy is simply:

```
objcopy -I input-format -O output-format infile
    outfile
```

where *input-format* is the input filetype (this can usually be omitted; objcopy can autodetect the format of the input files you will generally be working with, in the main), *output-format* is the output filetype (see below), *infile* is the name of the input file, and *outfile* is the name of the desired output file.

Like most other GNU utilities that work on executable file formats, objcopy uses the BFD (Binary File Descriptor) libraries to perform input and output formatting. If you invoke objcopy -?, the last line of on-line help will be a list of supported input and output filetypes. For example, the version of arm-elf-objcopy you compiled earlier will report:

```
arm-elf-objcopy: supported targets: elf32-
littlearm elf32-bigarm elf32-little elf32-big
srec symbolsrec tekhex binary ihex
```

Of these filetypes, probably the three that will interest you the most are:

- srec (Motorola S-record, a text file format commonly used for flash upgrade files and recognized by most PROM burning software),

- ihex (Intel HEX format, again recognized by most PROM burning software), and

- binary (raw binary format).

So (for example), to convert your program myprog.elf to a raw binary dump suitable for loading into flash memory, you can use the command line:

```
arm-elf-objcopy -O binary myprog.elf myprog.bin
```

objcopy also supports a large number of options (some of which are specific to particular input and output formats) for stripping particular information from the file, or for providing special instructions about how to convert information that is not represented unambiguously. It's fairly rare to use these options in an embedded project, particularly since many of these options control details that you should have set up correctly at link time. However, if you find that you need something more than simple "dumb" file conversion, you should refer to the on-line documentation for objcopy. For instance, you can take an ELF executable and (in two passes) copy the .text section to one output file and the .data section to a different file. This could be useful, for instance, if these two sections are to be loaded into separate memory devices.

Objdump—Check Your Executable's Layout

Objdump is an invaluable utility for "sanity-checking" your code. You will use it in particular when debugging issues related to linker scripts and memory maps. The general syntax is objdump [options] executable, where *executable* is the name of the file you want to inspect, and *options* specifies the information you want to receive. You can view a list of supported options using objdump —help, but I will describe some of the more useful options here. Most of the options have a short, single-character form and a longer "friendly" name; you can use these interchangeably.

--debugging

Dumps debugging information out of the specified file. This is mostly information that associates specific line numbers of sourcecode files with particular VMAs in the executable, and it isn't often going to be very useful to you. When run over the first code example below, arm-elf-objdump —debugging yields the following output:

```
example1.elf:       file format elf32-littlearm
boot.s:
/* file boot.s line 17 addr 0x2000000 */
/* file boot.s line 18 addr 0x2000004 */
/* file boot.s line 19 addr 0x2000008 */
/* file boot.s line 20 addr 0x200000c */
/* file boot.s line 21 addr 0x2000010 */
/* file boot.s line 22 addr 0x2000014 */
/* file boot.s line 23 addr 0x2000018 */
/* file boot.s line 24 addr 0x200001c */
/* file boot.s line 29 addr 0x2000020 */
/* file boot.s line 30 addr 0x2000024 */
/* file boot.s line 32 addr 0x2000028 */
/* file boot.s line 33 addr 0x200002c */
/* file boot.s line 36 addr 0x2000030 */
/* file boot.s line 37 addr 0x2000034 */
/* file boot.s line 38 addr 0x2000038 */
/* file boot.s line 39 addr 0x200003c */
/* file boot.s line 40 addr 0x2000040 */
/* file boot.s line 42 addr 0x2000044 */
/* file boot.s line 45 addr 0x2000048 */
/* file boot.s line 46 addr 0x200004c */
/* file boot.s line 47 addr 0x2000050 */
/* file boot.s line 48 addr 0x2000054 */
/* file boot.s line 51 addr 0x2000058 */
```

--demangle=*style*

If you specify the --demangle option, objdump will attempt to demangle symbol names; particularly useful when inspecting C++ code. The optional *style* parameter specifies what demangling style to use.

-d or --disassemble
-D or --disassemble-all

All four of these directives generate a disassembly of the specified input file. The -d and --disassemble directives only disassemble sections that are believed to contain code; the -D and --disassemble-all directives generate disassembly for everything in the file, regardless of whether or not it normally contains code bytes.

You can modify the disassembly output format using the options --prefix-addresses (which prints the complete address on each disassembled line) and/or --disassemble-zeroes, which forces objdump to disassemble blocks of zero-filled memory as code. If the file you are inspecting contains debugging information, you can also add filename and sourcecode line numbers using the --line-numbers option.

-h, --header or --section-header

Dumps all section headers. When debugging linker scripts, you will make frequent use of this option. For example, arm-elf-objdump -h yields the following output when run on the first example program in the next chapter:

```
example1.elf:      file format elf32-little

Sections:
Idx Name      Size      VMA       LMA       File off  Algn
  0 .text     00000068  02000000  02000000  00008000  2**2
                CONTENTS, ALLOC, LOAD, READONLY, CODE
  1 .data     00000000  02000068  02000068  00008068  2**0
                CONTENTS, ALLOC, LOAD, DATA
  2 .bss      00000000  02000068  02000068  00008068  2**0
                ALLOC
  3 .stab     0000012c  00000000  00000000  00008068  2**2
                CONTENTS, READONLY, DEBUGGING
  4 .stabstr  00000008  00000000  00000000  00008194  2**0
                CONTENTS, READONLY, DEBUGGING
```

As you can see, the .text segment contains the entire program and data. It is loaded at address 0x02000000, and is not relocated. (By the way, the legend underneath each line indicates the section flags. Refer to the section headed "Code Sections and Section Directives" above for more information on this topic.) The Algn (alignment) field indicates how the segment is to be aligned in memory, as a power of two.

-r or --reloc

Display relocation information for the executable. Unless you are developing for a specific operating system, your program will probably not contain relocation entries. You can, however, see relocation entries in the constituent object modules (.o files) of your projects. This information is not particularly valuable for the embedded developer.

--section=*section_name*

Only output information for the specified section named *section_name*.

--start-address=*address* and --stop-address=*address*

These two options allow you to specify a range of addresses for the **-d**, **-r** or **-s** options.

-t or --syms

Dump symbol table. This is another function of objdump which can help you sniff out and rectify problems with your linker script. Arm-elf-objdump -t example1.elf yields the following information from the first code example:

```
example1.elf:        file format elf32-littlearm

SYMBOL TABLE:
02000000 l    d  .text    00000000
02000068 l    d  .data    00000000
02000068 l    d  .bss     00000000
00000000 l    d  .stab    00000000
00000000 l    d  .stabstr 00000000
00000000 l    d  *ABS*    00000000
00000000 l    d  *ABS*    00000000
00000000 l    d  *ABS*    00000000
02000020 l       .text    00000000 reset
0200005c l       .text    00000000 PIO_SODR
02000060 l       .text    00000000 PIO_CODR
0200002c l       .text    00000000 blink_loop
02000064 l       .text    00000000 delay_constant
02000038 l       .text    00000000 pause_loop1
```

```
0200004c l    .text    00000000 pause_loop2
02000068 g    .data    00000000 __data_start__
02000068 g    .data    00000000 edata
02000068 g    .bss     00000000 __bss_start__
02000068 g    .data    00000000 datastart
02000068 g    *ABS*    00000000 end
02000068 g    .data    00000000 __data_end__
02000068 g    .text    00000000 etext
02000068 g    .bss     00000000 __bss_end__
02000068 g    *ABS*    00000000 __end__
02000068 g    .data    00000000 _edata
02000068 g    *ABS*    00000000 _end
02000000 g    .text    00000000 vectors
```

-x or --all-header

Displays all header information, including symbol table and and relocation entries. This is equivalent to specifying -a -f -h -r -t.

Size—Check the Load Size of Your Executable

Size is a simple little utility that totals up the size of the text, data and bss sections of your program. You invoke it simply as size *myprogram*. For example, the output given by arm-elf-size for the first example program in the next chapter is:

```
text    data    bss    dec    hex filename
 104       0      0    104     68 example1.elf
```

One main use for this utility is to determine quickly how much of your executable file is actual code and data, and how much is extraneous (such as debugger-related information). Another use is to check how much space is being occupied by the various sections, for the purpose of RAM or flash memory budgeting. There's nothing size will tell you that objdump won't, but size is simpler to use and provides a quick, easy-to-read result.

Gdb—The GNU Debugger

Gdb is your porthole into the target processor's environment. It is a software-only product on the host end that, for embedded targets, communicates with a back-end over serial, Ethernet or some other communications channel. This back-end is usually a snippet of software called "gdb stubs" (sourcecode for which can be found included in the standard gdb source distribution you have on the included CD-ROM). However, gdb isn't limited to talking its own language; it also supports various standardized debugger protocols used by other manufacturers, such as the Angel ROM monitor provided by ARM. If your board has no ROM monitor or debugger at all, then you can use an external piece of hardware that connects to a hardware debugging port and translates the hardware debugging protocol into a format understood by gdb. An example of such hardware is the Macraigor Systems Raven, which connects a JTAG/ICE interface on your target board to an Ethernet network for very high-speed, gdb-compatible debugging. If you don't have the budget for such hardware, or if your target doesn't have a supported hardware debugging interface, then your best route to connect it to gdb is to port the gdb stubs to run on your hardware.

Note that the use of bare gdb stubs is in some instances gradually being supplanted by the use of Red Hat's RedBoot loader/debugger back-end module. If you have to port one or the other, it's worth investigating the extra effort to get RedBoot up and running, because it provides many more services than simple debugging stubs.

Gdb also allows you to debug a program that is running on your own system. You are only likely to use this capability if your end project runs on a desktop version of Linux or NetBSD, and you are developing directly on the target hardware.

By the way, you will observe that the version of gdb we're using is a text-based application. There does exist a more attractive graphical version of gdb called Insight, developed by Red Hat. The reason we're using the text version here is that Insight has historically had serious problems running under Cygwin on Windows 95 and 98, and I wanted this text to be as general-purpose as possible. There's nothing you can do with Insight that you can't do with gdb, in any case; the graphical version is just more pleasing to the eye, and protects you (to a slight degree) from having to learn all of gdb's commands. If you want to experiment with Insight, you can obtain it from sources.redhat.com/

insight—the build process is exactly the same as for gdb. All of the example sourcecode and gdb techniques mentioned in this book will work with Insight without requiring any special modifications—Insight is in fact just a graphical wrapper around the same debugger code.

Invoking and Quitting gdb and Loading Your Program

To invoke the version of gdb or Insight you built using the instructions earlier in this book, assuming that the /tools/arm-elf/ bin directory is already in your PATH, simply use the command:

```
arm-elf-gdb options executable
```

where *executable* is the name of the ELF program file to be debugged, and *options* specifies zero or more additional command-line options (described below).

On startup, unless you have suppressed initialization scripts with the -nx parameter, gdb will first look in your home directory for a file called .gdbinit[20]. If this is found, it will be read and executed just as if you had typed its contents on the gdb command line. Gdb will then process any command-line parameters. Finally, it will look in the current directory for a .gdbinit file. Note how any settings that are altered during any of these steps may be overridden by a later step.

There are many additional command-line options you can specify, but most of them are not relevant to debugging embedded systems. Here is a brief summary of the options that are likely to be of interest to you:

-b *baudrate* or -baud *baudrate*

Use *baudrate* as the communications speed when using a serial debugging link.

---

[20] Some special versions of gdb use a different name for the initialization file. The information presented here is correct for the gdb version you built following the instructions in this book.

-command *filename* or -c *filename*

Execute *filename* as a command script.

-d *dirname* or -directory *dirname*

Add *dirname* to the path searched when looking for sourcecode files. This is useful when your project involves several assorted modules (such as third-party libraries). Specifying a complete set of search paths allows gdb to provide line-by-line symbolic debugging of the library code as well as your own.

-n or -nx

Do not process the default initialization files.

-symbols *filename* or -s *filename*

Read symbol-table information from *filename* instead of the main executable.

-t *device* or -tty *device*

Use *device* as the I/O device for serial-based debugging. For example, -t /dev/ttyS0.

When gdb has finished starting up, it will present you with a (gdb) command prompt and await your instructions. At this point, unless you specified some options on the command line or in some initialization script, gdb is *not* connected to the target. Your next steps would usually be in this order:

1. Configure any link parameters required to communicate successfully with the target. For example, for serial targets you would issue a command like `set remotebaud 9600` to set up the debugger link baud rate.

2. Tell gdb where to find the target hardware, with a command like `target rdi /dev/ttyS0`. This step also opens the debugging connection to the target board. In most cases, this will halt any code that is already executing on the target.

3. Tell gdb to send your program file to the target, with the `load` command.

4. Optionally set some breakpoints or watchpoints, if you want execution to halt at some particular event.

5. Commence program execution.

At some later moment, you might want to inspect the program's state. You can do this by pressing Ctrl-C to halt execution and return to a (gdb) prompt.

Whether program execution stops because of an exception, because of hitting a breakpoint or watchpoint, or because of your explicit intervention, gdb offers you a rich and varied set of commands to inspect and modify the target environment. A subset of these commands are documented in the remainder of this section.

When you Ctrl-C or otherwise halt a program, you can rebuild it without leaving gdb simply by typing make (followed by any required build parameters, such as make mytarget2) at the gdb prompt. If you need to run any command other than make, you should use the gdb command shell *commandline*, which simply runs whatever is in *commandline* as if you had opened a shell and typed it.

You can close the connection to your target device without quitting gdb by using the detach command. This can be useful if you have two target systems attached to the same host PC and you want to switch rapidly between them.

When you have finished with your gdb session altogether, issue the quit command (which can be abbreviated simply 'q'). gdb will detach from the target and exit, perhaps warning you in the process that the target program is still running.

Note: A variety of conditions can cause gdb to hang, particularly when debugging an RDI (Angel) target over a serial link. If this occurs, resetting the target may clear the hang condition. You may, however, end up in a situation where pressing reset on the target causes gdb to show the ROM monitor sign-on text, but gdb itself is unresponsive to your input. If this happens, you will need to use your host operating system's method for terminating crashed processes (in Linux, for instance, killall arm-elf-gdb will do the job).

Examining Target Memory

Using gdb, you can inspect arbitrary memory areas on the target using the x (eXamine) command. The general format of this command is:

```
x/REPEATformatSIZE start-address
```

(The reason I have mixed the case of the parameters above is simply for editorial clarity. There is no spacing between these parameters on the gdb command line, and I needed to make it clear that there are three distinct parameters immediately following the slash character.)

The optional *REPEAT* parameter is the number of items to be displayed. If this parameter is not specified, then 1 is assumed.

Format is a letter that specifies the way to format displayed data in the selected memory region. It can be any one of the following:

a	address
c	character
d	decimal
f	floating-point number
i	machine-language instruction
o	octal
s	string
t	binary
u	unsigned decimal
x	hexadecimal

Size is another optional single-letter parameter specifying to gdb the unit size of the item being dumped. It can be any of the following:

b	byte
h	16-bit halfword
w	32-bit word
g	64-bit giant

If *format* or *size* are not specified, then the last values supplied for these parameters are used. You can see some example usage of the x command in the chapter on worked firmware examples.

Breakpoints and Other Conditional Breaks

Gdb supports three ways of automatically breaking into your program when certain events occur; breakpoints, watchpoints and catchpoints. Catchpoints are not of quite so much interest to the embedded developer; they are triggered when a C++ exception is thrown, or when certain other events such as dynamic library loading occur. On the other hand, breakpoints and watchpoints are of great interest. A breakpoint is triggered when the program execution reaches some specified point, usually a function entry point or code label. A watchpoint is triggered when the value of some expression changes.

You set a breakpoint using the `break` command (this can be abbreviated 'b'). There are several variants of this command:

break *function-name*

Set a breakpoint on the start of function *function-name*. (You can also specify the name of an assembly-language symbol, though this is not strictly a function name).

break *+offset* or break *-offset*

Set a breakpoint at a given *offset* (in sourcecode lines) from the current program location.

break *line-number*

Set a breakpoint at the given line number *line-number* in the current sourcecode file.

break *filename:line-number*

Set a breakpoint at the given line number of sourcecode file *filename*.

break *memory-address

Set a breakpoint at an arbitrary memory location (even if this location is not occupied by your code).

break

Set a breakpoint at the address of the next instruction to be executed.

break *breakpoint-location* if *condition*

Set a breakpoint at *breakpoint-location* (using one of the syntaxes above) that will be triggered only if *condition* is true. For example, break myfunc if counter=0.

tbreak *breakpoint-location*

Set a one-time breakpoint at *breakpoint-location* (using one of the syntaxes above). This breakpoint is automatically removed after the first hit.

An exactly similar syntax can be used to set hardware-based breakpoints (using hbreak and thbreak commands), but many target platforms do not have hardware support for such breakpoints.

Watchpoints can be set in a similar manner, using the following commands:

watch *expression*

Gdb will break when *expression* is written with a new value. It will not break if expression is written but the contents are unchanged.

rwatch *expression*

Gdb will break when the program reads *expression*.

awatch *expression*

Gdb will break when the program reads or writes *expression*, regardless of whether the value is changed or not (note the subtle difference between this and the watch command).

Each time you assign a breakpoint, watchpoint or catchpoint, gdb assigns it a number. You use this number to refer to the breakpoint (etc.) when enabling, disabling or removing it. You can get information about currently set breakpoints, watchpoints and catchpoints, including status information and statistics on the number of hits, by using the commands `info breakpoints` and `info watchpoints`.

You can delete a breakpoint or range of breakpoints (or watchpoints and catchpoints) with the `delete n` command, where *n* is a specific breakpoint number or range of numbers, e.g., `delete 1-8`. If you do not specify a parameter, gdb will ask for confirmation and then remove all breakpoints, watchpoints and catchpoints.

Getting Further Help

The above descriptions and command definitions are a useful minimum introduction to gdb's capabilities, and in my experience embedded programmers often go no further than the above functionality. For further information on more advanced functions, you need look no further than the program itself: gdb has an extremely large on-line help database. Typing `help` by itself will show you some top-level categories; `help command` will give you detailed help on a specific *command*, or a specific category of commands (e.g., `help breakpoints`). Gdb also has an info page, of course, and I encourage you to refer to it if you are curious about the other capabilities of this utility.

## Chapter 4

Example Firmware Walkthroughs and Debugging Techniques

A Quick Introduction to ARM and the Atmel EB40

Although I promised at the start of this book that you wouldn't need specific experience with the ARM architecture, in order for the code examples in this chapter to be fully comprehensible, it is useful to have a grasp of the the ARM architecture in general, and the Atmel EB40 in particular. If you're already familiar with this board, or if you simply want to peruse the code in module-functionality terms and don't need a detailed explanation of the board's workings, feel free to skip this section.

If you intend to compile the examples and test them on a real board, you should make sure that you have the following hardware items:

- Atmel EB40 board.

- Two available serial ports on your computer (or two computers).

- One straight-through serial cable with a male DB9 connector on one end and a suitable connector (female DB9 or DB25) to fit your computer on the other end.

- One nullmodem cable with a female DB9 connector on one end and a suitable connector (female DB9 or DB25) to fit your computer on the other end.

- Simple ASCII terminal emulation software. Hyperterminal under Windows, or minicom under Linux are both suitable; no special features are required.

- GNU tools built for your operating system and the ARM target, as described in an earlier chapter.

Please note that the information in this section is a *very* light introduction to the ARM core, and I am intentionally not going into great detail about the ARM programming model and instruction set; I am only discussing features to which I make specific later reference in code examples. The definitive guide and essential reading on this topic is the ARM7TDMI datasheet (document number ARM DDI 0029E). This is part of the documentation portfolio for the AT91R40807 microcontroller, and can be downloaded from Atmel's web site. At the same time, you should download the AT91M40800/AT91R40807/AT91M40807/AT91R40008 datasheet, of which the current version at the time of writing is Rev. 1354D-ATARM-05/02. This latter document provides a detailed description of the microcontroller's peripherals and other important features.

The EB40 and the AT91R40807 microcontroller it contains are relatively simple devices in the spectrum of ARM parts. The core is a typical von Neumann architecture, with a single address space (all I/O is memory-mapped). The first 32 bytes of memory are reserved for the exception vector table:

Location	*Function*
0x00000000	Power-on reset
0x00000004	Undefined instruction
0x00000008	Software interrupt
0x0000000C	Prefetch abort
0x00000010	Data abort
0x00000014	Reserved
0x00000018	IRQ (interrupt request)
0x0000001C	FIQ (fast interrupt request)

Note that these locations do not store jump target addresses, as in architectures like the MC680x0; the table stores actual *jump instructions*. The processor performs an absolute jump to the appropriate location when the relevant exception occurs. Because of this, the ARM needs to have valid boot code at 0x00000000 on power-up and hence this location needs to be mapped to ROM, at least initially. However, it is advantageous for software to be able to modify the exception vector table. Partly for this reason, almost all ARM variants either have a memory-management unit (MMU) that can be used to remap RAM to logical address 0x00000000, or some internal chip selection logic that can be used to alter the mapping of physical addresses to external chip select lines sometime after initial boot. The Atmel AT91R40807 has the latter scheme; a simple programmable address selection scheme accessed via special function registers. Like most other ARM parts, it has built-in address decoding logic that generates external chip select signals during external bus read/write operations. These signals are named NCS0 through NCS3 and CS4 through CS7 (the latter four are multiplexed with address lines, and are active-high).

Thus, there are several different memory maps you will need to consider for the EB40. Firstly, let's look at the memory layout of the board at power-on[21]:

Start	*End*	*Contents*
0x00000000	0x0001FFFF	128K flash (NCS0)
0x00100000	0x0011FFFF	128K on-chip SRAM
0x00300000	0x00301FFF	8K on-chip SRAM
0xFFC00000	0xFFFFFFFF	Peripherals

The flash memory is divided by external hardware into two sections: The lower 64K contains the Atmel bootloader and Angel debugger software, and it is write-protected (although it can be write-enabled by fitting a jumper to the board). The upper half of the chip is available for user programs, and it is factory-loaded with a small demo program. This upper 64K area can be moved to location 0x00000000 by means of a switch that ties A16 either to the A16 line from the CPU (in which case the memory

---

21 Because of address decoding limitations, all of the items in the EB40's memory map are mirrored at several locations. For clarity, the mirrored copies are not shown.

map is as above), or directly to Vcc (in which case the upper 64K half of the flash memory appears mirrored in the address space at locations 0x00000000 and 0x00010000). Using this system, you can boot the board using Angel, load a flash-loader into RAM, and burn your own program into the upper half of flash memory. You can then flip the switch, reset the board, and your own program will run directly out of flash.

Note: Due to a bug (or rather an omission) in the EB40's hardware design, it is not possible to enter flash programming mode when the write-protect jumper is open (i.e., in the protected mode). In order to write the user area of the flash chip, you have to unprotect the entire device by fitting a shorting block on jumper J7, contrary to what the Atmel documentation tells you. **Be very cautious about writing to flash memory on this board**. If, for instance, your flash-writer routine has a bug that accidentally writes too much data to the user area, the address pointer will wrap around to the bottom of the flash device (due to partial address decoding on the EB40) and you will obliterate the bootloader and Angel. Once this happens, your *only* route to get code back into the board is with a JTAG-based debugger module. It would have been vastly preferable if the write-protect jumper worked the way Atmel had intended it to work, but there doesn't seem to be a way to achieve this original design goal that doesn't require quite complicated external logic.

For insatiably curious readers, the details of the EB40's hardware design bug are as follows: The EB40 gates the write enable signal running to the flash chip with the CPU's A16 output line and jumper J7. If J7 is open-circuit, the write enable signal to the AT29LV1024 flash chip will only be asserted if the A16 output from the CPU is high. In other words, the chip is effectively write-protected for addresses in the lower 64K of flash memory. This sounds more or less exactly like the design goal Atmel had in mind. However, in order to perform a sector write operation on the AT29LV1024, you need to issue a special unlocking sequence of write cycles. This mechanism protects the chip against accidental corruption due to transient signals (especially during powerup and powerdown.) The unlocking sequence is: Write 0xAAAA to flash address 0x5555, then write 0x5555 to address

0x2AAA, then write 0xA0A0 to address 0x5555. (Note that these are *word* addresses; i.e., they are the binary pattern the flash chip expects to see on its address bus. Because of the way the part is wired on the EB40, the actual addresses to be written, from the CPU's perspective, are shifted left one bit). Unfortunately, these special unlock addresses are in the lower 64K of flash. The EB40 has no way of knowing that these write cycles are simply unlocking cycles, so they are blocked unless J7 is shorted. The net effect is that the entire flash chip is write-protected by J7.

Assuming that you have the switch in its "LOWER MEM" position, with A16 under CPU control, Angel will set up the following memory map:

Start	End	Contents
0x00000000	0x00001FFF	8K on-chip SRAM
0x00100000	0x0011FFFF	128K on-chip SRAM
0x01000000	0x0101FFFF	128K flash (NCS0)
0x02000000	0x0203FFFF	512K SRAM (NCS1)
0xFFC00000	0xFFFFFFFF	Peripherals

The third memory map you need to consider is the arrangement that will be in place when running your own code directly out of flash memory; and to a certain degree, this is under your control. The AT91R40807 has two modes of operation; a simple power-on boot mode where all external memory accesses are pointed to the memory device on chip select NCS0 (the memory map for this is illustrated above), and a more complex mode where the 8K block of internal SRAM is mapped at 0x00000000, the internal 128K SRAM block is mapped at 0x00100000. In this mode, NCS0-3 (and, optionally, CS4-7) are all asserted in different blocks of the memory map. The idea is that you set up the chip select mode registers at boot time, copy your exception vector table to the 8K SRAM block, and probably also move your main code to the 128K on-chip SRAM (in some other variants of this processor family, that 128K area is occupied by ROM). You then jump to the code's new location and execute a "remap" command to enable the external chip selects.

First Step—the LED Flasher (in Assembler)

The day has finally come—after weeks of poring over datasheets and drawing schematics, you finally have the first run of prototype PCBs on your bench, and it's time to get them working and port your firmware. Of course, we know the Atmel EB40 works, but for educational purposes we will treat it as if it were our own, largely unverified hardware platform.

The first thing we're going to do is write a very small assembly-language program to check that the processor is running, and that the interconnects to RAM and ROM are correct. This program will blink one of the EB40's LEDs rapidly; the embedded systems equivalent of "Hello, world!". (When I'm writing this kind of quick and dirty test code, I usually just put an oscilloscope on one of the output pins, so I don't have to spend any time thinking about drive current capabilities or timing constants for the flash loop. However, the EB40 has a convenient set of LEDs ready-wired to appropriate GPIO pins, so we may as well use them.)

The sourcecode for this program can be found on the CD-ROM in the directory sourcecode/blink. To build it, copy it to a temporary directory on your hard drive, change to that directory, and run make[22].

Note that the EB40 gives us at least five ways to load and execute this code. We can use gdb to load the code into RAM using the Angel loader, we can use the EB40's proprietary bootloader to achieve the same thing, we can poke it directly into RAM using a JTAG module, or we could use either the JTAG interface or our own custom flash-writer code to write our program into the unoccupied 64K of flash memory on the board.

For the time being, however, we will use gdb. One reason for choosing gdb over the EB40's bootloader, even for trivial applications like this, is that the Atmel BINCOM utility, required to load code directly to RAM using the bootloader, is only provided

---

[22] If you use Windows to copy the files from a CD-ROM, the copies on your hard drive will have their read-only flag set; remember to make the files writable using ATTRIB or by right-clicking each file in Explorer and unchecking "Read-only".

for Windows, whereas everything else I am discussing in this chapter is equally valid for Linux or Windows.

Ensure that you have a straight-through serial cable connected from the SERIAL A port on the EB40 to your PC, and that the EB40's memory switch SW1 is set to "LOWER MEM". Connect power to the EB40, press the red reset button and you should see the yellow LED (LED2) illuminate, indicating that Angel is running. Now open a shell prompt, change to the directory containing your local copy of the example1 sourcecode, run make to build the executable, then type the following commands[23]:

```
arm-elf-gdb example1.elf

set remotebaud 9600

target rdi /dev/ttyS0

load
```

(I've assumed above that the EB40 board is connected to the PC via the first serial port, which is normally /dev/ttyS0. If your EB40 is connected to a different port, make the appropriate substitution. Also note that although Angel does support higher transfer speeds, it's much more reliable when you run at 9600bps. Feel free to experiment with speeds as high as 115200bps, but be prepared for the debugger to hang at any speed other than 9600bps. The typical symptom is that you will see the Angel sign-on message, then gdb will be hung hard; you'll need to kill the process.)

The RDI target specified in the third command line stands for Remote Debug Interface; it is ARM's name for the serial protocol used by Angel. If you were connecting to a target that has gdb stubs running on it, you would use the command `target remote /dev/ttyS0` instead.

To start the program, type `continue` and press Enter. The red LED on the board will start blinking at a rate of about 2Hz.

Let's leave the hardware for the time being and examine the program in detail. First, we'll look at the Makefile:

---

[23] Remember to make sure that your ARM tools are in the current PATH, as detailed in the chapter on installing the toolchain.

```
# Embedded Systems Development on a Shoestring
# Example project 1 - EB40 LED Flasher
#
# Lewin A.R.W. Edwards (sysadm@zws.com), Jun-2002.

ASFLAGS = -mcpu=arm7tdmi -gstabs
LDFLAGS = -Teb40-ram.ld -nostartfiles -Lgcc -L.
OBJS    = boot.o
EXE     = example1.elf

$(EXE): $(OBJS)
        arm-elf-gcc $(LDFLAGS) -o $(EXE) $(OBJS)

%.o:%.s

        arm-elf-as $(ASFLAGS) $< -o $@

clean:
        rm -f $(OBJS)
        rm -f $(EXE)
```

There isn't a lot of complexity here. We are generating an executable file called example1.elf, which is composed of a single object file, boot.o. We also provide a simple rule to generate this object file from an assembly sourcecode file. You could use this makefile as a generic template for writing all-assembly-language programs.

The -gstabs parameter given to gas tells it to generate and include symbol table information in .stabs and .stabstr sections. This will be helpful in later debugging (though the structure of this first ultra-simple example program is such that it doesn't actually generate any debugging information in this section).

We can now inspect the actual sourcecode to the program:

```
@@@@@@@@@@@@@@@@@@@@@@@@@@@@@@@@@@@@@@@@@@@@@@@@@@@@@@@@@@@@@@@@
@ Embedded Systems Development on a Shoestring
@
@ Example project 1 - EB40 LED Flasher
@
@ Lewin A.R.W. Edwards (sysadm@zws.com), Jun-2002.
.section .text
.code 32
.globl vectors
```

```
@@@@@@@@@@@@@@@@@@@@@@@@@@@@@@@@@@@@@@@@@@@@@@@@@@@@@@@@@
@ In a ROM-startup program, this would be the interrupt
@ vector area. As this program is intended for RAM
@ startup, we simply have our init code at the "reset
@ handler" (entry point).

vectors:
        b reset         @ Reset
        b .             @ Undefined instruction
        b .             @ SWI
        b .             @ Prefetch abort
        b .             @ Data abort
        b .             @ reserved vector
        b .             @ irqs
        b .             @ fast irqs

@@@@@@@@@@@@@@@@@@@@@@@@@@@@@@@@@@@@@@@@@@@@@@@@@@@@@@@@@
@ Entry-point code
reset:
        ldr r4, PIO_SODR
        ldr r5, PIO_CODR

        @ Bit 1 is the red LED on the EB40
        ldr r6, =0x02
blink_loop:
        str r6, [r4]  @ Turn on LED

                        @ Brief pause
        ldr r0, delay_constant
        ldr r1, =0
pause_loop1:
        sub r0, r0, #1
        cmp r0, r1
        bne pause_loop1

str r6, [r5]            @ Turn off LED

                        @ Brief pause
        ldr r0, delay_constant
pause_loop2:
        sub r0, r0, #1
        cmp r0, r1
        bne pause_loop2
```

```
                              @ Loop forever
          b blink_loop

@@@@@@@@@@@@@@@@@@@@@@@@@@@@@@@@@@@@@@@@@@@@@@@@@@@@@@@@@@@@@@@
@ Miscellaneous constants
                              @ Set Output Data Register
PIO_SODR: .word 0xffff0030
                              @ Clear Output Data Register
PIO_CODR: .word 0xffff0034
                              @ Provides a decent delay period
delay_constant:              .word 0x00040000
```

Since this code uses only registers to store its loop variables and other temporary data, everything lives in the .text section, hence we have a single .section directive at the start of the file to indicate this fact.

What's this ".code 32" directive, though? This directive is specific to the ARM targeted version of as. Many variants of the ARM core have two modes; normal ARM mode, with a 32-bit instruction word, and a cut-down mode called Thumb mode. In Thumb mode, the processor operates with a limited instruction set, fewer directly accessible registers, and a 16-bit instruction word. Thumb code has two main uses—firstly, it's more space-efficient, and secondly, it is more bandwidth-efficient on systems with a narrow code memory bus. In fact, the AT91R40807 is just such a processor, because it has a 16-bit external data bus; it's significantly more efficient to program this device using the Thumb extension.

For simplicity, however, I want to work through the code examples in this book with the full 32-bit ARM instruction set, hence I use the .code 32 directive to inform gas to assemble 32-bit opcodes. This is really the native instruction set of the ARM core—not all ARM cores support Thumb, and once you start mixing Thumb and ARM code in your program, you are forced to consider various processor mode switching and context saving tasks that aren't necessary in a pure-ARM program. (It's not possible to write a pure Thumb program, because different events—power on reset and various exception conditions—can force the processor back into ARM mode.)

If you want to experiment with Thumb code for yourself, you can coerce gcc to generate Thumb opcodes by supplying the -mthumb switch on its command line. You'll probably want to add the -mthumb-interwork switch also, which includes extra preamble code allowing Thumb functions to call ARM functions and vice versa (this is referred to as "Thumb interwork code"). Please note that you'll also need to rebuild newlib, because the version we compiled earlier is built for 32-bit ARM only, without interworking support.

Note that I have called the entry-point "vectors" and the start of this code is a fully-populated ARM exception vector table. Although this is in no way necessary for a program that will be loaded into an arbitrary RAM location, it does pave the way for a future step of migrating the code to boot directly out of flash memory.

The functional part of the code is fairly self-explanatory. We know that the Atmel bootloader has already set up the necessary data direction registers in the processor, so all we need to do in order to turn an LED on and off is set and clear the appropriate bit in the GPIO port register. On the AT91R40807, this is achieved by means of two registers; PIO_SODR and PIO_CODR. When you write to PIO_SODR, any '1' bit in the word you write is SET in the GPIO output data latch—other bits are undisturbed. When you write to PIO_CODR, any '1' bit in the word you write is CLEARED in the GPIO output data latch. If you wanted to load some specific word *gpioword* into the data latch, you could use code something like this:

```
PIO_SODR = gpioword;
PIO_CODR = gpioword ^ 0xffffffff;
```

We use the following simple linker script to link the application:

```
/*

Linker script for Atmel EB40

This script is intended for C programs
that will be loaded with BINCOM and the
Atmel bootloader, or via Angel. All
segments are emitted to the SRAM block
at 0x02000000.
```

```
        Lewin A.R.W. Edwards (sysadm@zws.com),
        Feb-2002
*/
/*

        For standardization, the entry-point is
        called "vectors", although for these programs,
        this is not actually the interrupt vector
        table.
*/

ENTRY(vectors)

SEARCH_DIR(.)

/*
        There is a single memory segment, representing
        the 512K on-board SRAM.
*/

MEMORY

{
        sram : org = 0x02000000, len = 0x00080000
        /* 512KBytes of SRAM */
}

SECTIONS

{
        .text :
        {
            *(.text);
            . = ALIGN(4);
            *(.glue_7t);
            . = ALIGN(4);
            *(.glue_7);
            . = ALIGN(4);
            etext  =  .;
        } > sram

        .data ADDR(.text) + SIZEOF(.text) :
        {
            datastart = .;
```

```
        __data_start__ = . ;
        *(.data)
        . = ALIGN(4);
        __data_end__ = . ;
        edata  =  .;
        _edata  =  .;
    }

    .bss ADDR(.data) + SIZEOF(.data) :
    {
        __bss_start__ = . ;
        *(.bss); *(COMMON)
        __bss_end__ = . ;
    }

    end = .;
    _end = .;
    __end__ = .;
    /* Symbols */
    .stab 0 (NOLOAD) :
    {
        [ .stab ]
    }

    .stabstr 0 (NOLOAD) :
    {
        [ .stabstr ]
    }

}
```

You'll observe that this script is much more complicated than
you might expect; this is because it's intended to be more general-purpose than this example program absolutely requires.

Note that the last two stanzas in this script refer to debugging
symbol table information. We want this to be emitted to the ELF
executable so that when we break the running program, gdb can
provide us with useful information about the current program
counter, cross-referenced to the relevant line in the appropriate
sourcecode file. Including this debugging information often makes
the ELF file much larger than the actual code contained in it, but

this isn't important; none of the extraneous information would be emitted to a ROM dump.

Speaking of debugging, it's time to see what gdb, in conjunction with Angel, can do for us. Press Control-C to quit the program running on the target board. Gdb will produce output something like this:

```
RDI_execute: you pressed Escape
Program received signal SIGINT, Interrupt.
pause_loop1 () at boot.s:40
40                              bne pause_loop1
Current language:  auto; currently asm
(gdb)
```

The exact output you see depends on just where the program was when you interrupted it. At this point, the program is frozen (you'll observe that the LED is no longer blinking). Since we don't know exactly where we are in the program when we stopped it, let's begin by setting a breakpoint at the start of the program's main loop. Type break blink_loop and press Enter, and gdb will report:

```
Breakpoint 1 at 0x200002c: file boot.s, line 33.
```

Resume program execution with the continue command, and almost immediately you will see the following output:

```
Continuing.

Breakpoint 1, blink_loop () at boot.s:33
33       blink_loop: str r6,[r4]    @ Turn on LED
(gdb)
```

Now the program on the target board is at a known location, we can delete the breakpoint we set earlier, by issuing a clear blink_loop command. We can now single-step through the program by issuing a step command.

Gdb supports a wide variety of breakpoint options. You can set a breakpoint on a label, using the syntax break *labelname*. You can also set a breakpoint to be triggered when execution reaches a specific line in a particular sourcecode file, using the syntax break *sourcefile:linenumber*. For more information, you can refer to gdb's comprehensive built-in help system; for example `help break` or `help clear`.

Bringing Up a Simple C Program— The LED Flasher (in C)

Now that we've verified that our board is fetching code and executing it correctly, we can start to bring up the C run-time library and hoist ourselves into a high-level application. Since we're loading our code into RAM and executing it directly from the load address, this is quite simple. (The situation gets much more complicated once we need to start up out of ROM.) All we need to do in order to pass control to the C program is to zero the BSS section, set up a stack pointer, and jump to the main program.

In this project, we'll use a variant of the linker script from the previous example. (All the sourcecode for this project can be found in sourcecode/blink_c on the included CD-ROM.) The only change we make to the linker script is to reserve some extra space (two kilobytes) in the .bss output section for the application's stack, and define a couple of symbols to let the main program know where the stack starts and finishes. The new .bss output section definition reads:

```
.bss ADDR(.data) + SIZEOF(.data) :
{
    __bss_start__ = . ;
    *(.bss); *(COMMON)
    __bss_end__ = . ;
    _stack_bottom = . ;
    . += 0x800 ;
    _stack_top = . ;
}
```

Notice, by the way, that I haven't initialized the stack area with any specific fill value. A more usual practice would be to fill the stack area with a distinctive magic value (0xDEADBEEF is one historically significant example of distinctive fill patterns). This can help you determine if you're over-allocating or under-allocating stack space; you can allow your program to run for a while (hopefully, putting it through some worst-case stack usage operations), and then inspect the stack area. The portion that still contains your magic value wasn't used and can be reclaimed. If the magic value no longer appears anywhere within the stack area, then your program almost certainly blew out its stack and overwrote some of whatever lay below the stack in memory.

In addition to updating the linker script, we also need to make a minor change to the Makefile; we add a new variable called CFLAGS that holds command-line switches to be passed to gcc, and we also need to add a new rule to compile .c sourcecode files. We also change the executable's name to example2.elf. Finally, we need to add a new module to the OBJS list; main.o, corresponding to our C sourcecode file main.c. Rather than reproduce the entire makefile, I'll just illustrate those changes here:

```
CFLAGS = -g -I. -mcpu=arm7tdmi

OBJS    = boot.o main.o
EXE     = example2.elf
%.o:%.c
        arm-elf-gcc -c $(CFLAGS) $< -o $@
```

The options on the gcc command line deserve a little explanation. The -g option tells gcc to generate debugging information in a gdb-friendly manner. -I tells gcc to search the current directory for include files. -mcpu=arm7tdmi tells gcc which particular variant of the ARM core we're using - the current version of gcc supports about thirty different ARM core variants from ARM2 to XScale. Note that the -c option (compile single file only; do not invoke the linker) is specified in the gcc rule, rather than in CFLAGS, because it is always going to be required in *any* project.

Now let's look at the assembly-language stub that prepares the C environment and loads the main program. Please note that I have omitted some of the comments in this sourcecode, because they duplicate information that is described completely in the previous example.

```
@@@@@@@@@@@@@@@@@@@@@@@@@@@@@@@@@@@@@@@@@@@@@@@@@@@@@@@@@@@@@
@ Embedded Systems Development on a Shoestring
@
@ Example project 2 - EB40 LED Flasher in C
@
```

```
@ Lewin A.R.W. Edwards (sysadm@zws.com), Jun-2002.

.section .text
.code 32
.globl vectors

vectors:
        b reset         @ Reset
        b .             @ Undefined instruction
        b .             @ SWI
        b .             @ Prefetch abort
        b .             @ Data abort
        b .             @ reserved vector
        b .             @ irqs
        b .             @ fast irqs

@@@@@@@@@@@@@@@@@@@@@@@@@@@@@@@@@@@@@@@@@@@@@@@@@@@@@@@@@@@@@@@@@
@ Entry-point code
reset:
        @ Begin by clearing out the .bss section
        ldr r1, bss_start
        ldr r2, bss_end
        ldr r3, =0

clear_bss:
        cmp r1,r2
        strne r3,[r1],#+4
        bne clear_bss

        @ Initialize the stack pointer
        ldr r13,stack_pointer

        @ Call main function
        bl main

        @ Loop forever if main() exits
        b vectors

@@@@@@@@@@@@@@@@@@@@@@@@@@@@@@@@@@@@@@@@@@@@@@@@@@@@@@@@@@@@@@@@@
@ Miscellaneous constants
stack_pointer: .word _stack_top
bss_start:     .word __bss_start__
bss_end:       .word __bss_end__

        .end
```

This is the simplest possible case of C run-time initialization. For instance, we don't provide any support for run-time library functions that may require extra board-specific code or symbol definitions. More importantly, our code has the rare quality of not requiring any special fixups or data relocation; it is loaded by gdb (or rather by Angel) directly to the area from which it needs to run. (In fact, the point is rather moot since we don't use any data or BSS space in this program, because the only variable we create is allocated on the stack. But even if we did create some initialized global variables, there's no data we would need to relocate—the initialized data segment is already in RAM at the correct VMA, and it is preloaded with its required initial values while the executable is being transferred onto the target board.)

By the way, note that register r13 is defined (as an ARM convention) to be the stack pointer, and gas assigns it an alias of "sp". Likewise, r14 is reserved for the link register used when returning from a subroutine, and gas allows you to refer to this register via the alias "lr", as in the oft-used "bx lr" return-from-subroutine code. Finally, r15 is the program counter, referred to by the alias "pc".

You won't often be able to use startup code this simple in an embedded application, because your program will almost always be running out of ROM; at the very least, you'll have to copy the initialized data segment from its LMA in ROM to its VMA in RAM. The one common scenario where you might be able to write your main application like this is if you have a bootloader on your board that performs all the loader functionality for you. For example, in order to reduce overall materials cost in a project, I once developed a system that had a very small bootloader in masked ROM (very cheap) and the application code in SmartMedia®-compatible NAND flash memory (again, very cheap—but not directly bootable, and very slow). The bootloader simply copied the NAND flash contents to the start of RAM and jumped into the main program thus loaded.

In that project, the application code in NAND flash used a linker script and initialization code much like the configuration in this little example program.

One important note applies here: Because we can, in general, assume that programs are going to make changes to their initialized variables (items in the .data section), programs that are linked and loaded using the above method way can't simply be restarted, because all variables in their initialized data sections will be corrupted with whatever the last information written by the program might have been. If you terminate execution of a program written this way, you need to reload the entire thing from the source media. In the case of a program being debugged using gdb, you need to execute the load command; simply issuing a run command to restart the target program will not necessarily work correctly.

Since we've now got the C run-time ready to roll, let's look at the C sourcecode for the meat of the program (this is the main.c sourcecode file):

```
/*

Embedded Systems Development on a Shoestring
Example project 2 - EB40 LED Flasher in C
Lewin A.R.W. Edwards (sysadm@zws.com), Jun-2002.
*/

/* GPIO set and reset registers */
#define PIO_SODR    0xffff0030
#define PIO_CODR    0xffff0034

/* Bits to be set/reset on the GPIO port in
   order to toggle the LED */
#define RED_LED    0x00000002

/* Quick and dirty macros to write a register
   or memory location word, halfword or byte */
#define WRITEREGW(addr,value) *((volatile
unsigned int *) (addr)) = (value)

#define WRITEREGH(addr,value) *((volatile
unsigned short *) (addr)) = (value)

#define WRITEREGB(addr,value) *((volatile
unsigned char *) (addr)) = (value)
```

```
int main(void)
{
        register int i;

        while (1) {
              for (i=0x40000;i>0;i--) {}
              WRITEREGW(PIO_SODR,RED_LED);
              for (i=0x40000;i>0;i--) {}
              WRITEREGW(PIO_CODR,RED_LED);
        }
}
```

This is fairly close to being a direct port of the assembly-language version. Once built with make, this program can be loaded and run on the board using the same method as we used for the first example, except that the filename for this second example is example2.elf.

If you load and run this second example the same way as the first, you will observe that even though I used the same nominal timing constant (0x40000) and algorithm in this project as in the previous project, the LED blink rate is much slower than it was in the pure-assembly language version; about 0.75Hz. As a matter of interest, you can improve the performance of this little application quite a lot by enabling gcc's optimization features. Simply edit the CFLAGS line and add the option -O3 (maximum optimization), rebuild the project with make clean ; make, load it onto the board and you'll see that the blink rate is about 35% faster.

Writing a Simple Flash-Loader (and Inspecting Memory with gdb)

The programs we've written so far all require an external entity—in our case, Angel—to load them into RAM ready for execution. It's time to demonstrate how we can remove our dependency on Angel by writing a program that can initialize itself from power-on. This is also the only way we can illustrate how to initialize more complex programs with relocated data segments, different VMA/LMA scenarios, and other exotica.

Before we can develop any such programs, however, we need to engineer a way to get them into the on-board memory. That is what we will do in this section. (In the next section, we'll develop a simple ROM-bootable program that can be used as a starting point for more complex and exciting programs.)

This example flash-loader program is unavoidably less generalizable than the previous examples; it's highly specific to the Atmel board. However, it serves as a useful illustration of an intermediate step you might take in developing your own board. You can find the sourcecode to this loader program in the sourcecode/flash directory on the included CD-ROM.

This program was developed with the sourcecode/blink_c code as a basis. Firstly, we add the following snippet to boot.s at the end (immediately before the .end directive):

```
@@@@@@@@@@@@@@@@@@@@@@@@@@@@@@@@@@@@@@@@@@@@@@@@@@@@@@@@@@@@@@
@ Data to be loaded into flash memory
.globl flash_contents
flash_contents:        .incbin "flash.bin"
                       . = flash_contents + 0x10000
```

This specifies that we want to include the external file "flash.bin", which will contain the raw binary data to be written into the user section of the EB40's flash memory. Explicitly changing the location counter after including the external binary file serves a couple of principally housekeeping purposes: Firstly, if your external program exceeds 64K in size (the limit of available flash memory, as you'll recall from our description of the EB40 hardware above), you'll be alerted to this fact by a fatal error when gas tries to move the location counter backwards. Secondly, although this isn't important on the EB40, if you were loading the flash-loader into a RAM area too large for it, you would get an error either at link time or at load time. It's always wise, where possible. to structure things so that your tools work as hard as possible at reporting errors of this nature—it can prevent a lot of fruitless debugging later. The one major downside to padding the program in this way is that the flash-loader module itself is automatically blown out to slightly more than 64K in size. That makes it somewhat slow to load over the 9600bps Angel serial link, but it's an acceptable compromise.

Next, we tweak the makefile a little. We change the output filename to flash.elf by altering the EXE variable:

```
EXE    = flash.elf
```

We also need to change the first rule to indicate that this project is now dependent on the flash.bin file:

```
$(EXE): $(OBJS) flash.bin
```

All of the major changes to the code are in the C section of the program, main.c, which you can read below:

```
/*

    Embedded Systems Development on a Shoestring
    EB40 Flash Programmer
    Lewin A.R.W. Edwards (sysadm@zws.com), Jun-2002.
*/

#include <string.h>

/* GPIO set and reset registers */
#define PIO_SODR    0xffff0030
#define PIO_CODR    0xffff0034

/* Bits to be set/reset on the GPIO port in
   order to toggle the LEDs */

#define RED_LED      0x00000002
#define YELLOW_LED   0x00000010
#define GREEN_LED    0x00000004

/* Angel-defined starting address of bottom of
   flash memory */
#define FLASH_BASE   0x01000000

/* Angel-defined starting address of writable
   area of flash memory */
#define FLASH_START FLASH_BASE + 0x10000

/* Number of words per sector in the AT29LV1024
   flash chip */
#define FLASH_PAGE_SIZE   128

/* Number of programmable bytes in the user-
   writable area of the flash chip */
#define FLASH_SIZE   0x00010000
```

```c
/* Quick and dirty macros to write a register
   or memory location word, halfword or byte */
#define WRITEREGW(addr,value) *((volatile
unsigned int *) (addr)) = (value)
#define WRITEREGH(addr,value) *((volatile
unsigned short *) (addr)) = (value)
#define WRITEREGB(addr,value) *((volatile
unsigned char *) (addr)) = (value)

/* Quick and dirty macros to read a register or
   memory location word, halfword or byte */
#define READREGW(addr) (*((volatile unsigned
int *) (addr)))
#define READREGH(addr) (*((volatile unsigned
short *) (addr)))
#define READREGB(addr) (*((volatile unsigned
char *) (addr)))

/* Import the flash-programming buffer from
boot.s */
extern unsigned short flash_contents[FLASH_SIZE / 2];

/*
    This subroutine loads the upper 64K of the
    flash chip. The red LED is on during the
    program operation. The yellow LED is on
    while waiting for the flash chip to finish
    a sector write. The green LED is turned on
    and the other LEDs turned off when the
    write is complete.
*/

void LoadFlash(unsigned short *contents)
{
    register int i,j;

    /* Turn off yellow and green LEDs */
    WRITEREGW(PIO_CODR, YELLOW_LED | GREEN_LED);

    /* Turn on red LED to indicate burn in
       progress */
    WRITEREGW(PIO_SODR, RED_LED);
```

```
        for (i=0; i<FLASH_SIZE; i+=FLASH_PAGE_SIZE * 2)
        {
            /* Unlock chip for sector programming */
            WRITEREGH(FLASH_BASE + (0x5555 << 1),
                0xaaaa);
            WRITEREGH(FLASH_BASE + (0x2aaa << 1),
                0x5555);
            WRITEREGH(FLASH_BASE + (0x5555 << 1),
                0xa0a0);

            for (j=0;j<FLASH_PAGE_SIZE;j++) {
                WRITEREGH(FLASH_START + i + (j * 2),
                    contents[(i/2)+j]);
            }
        /* Turn on yellow LED to indicate
            waiting for program finish */
        WRITEREGW(PIO_SODR, YELLOW_LED);

        /* Wait for sector to finish programming */
        while (READREGH(FLASH_START) !=
            READREGH(FLASH_START)) { }
        /* Turn off yellow LED */
        WRITEREGW(PIO_CODR, YELLOW_LED);
        }
        /* Burn complete - turn off red LED */
        WRITEREGW(PIO_CODR, RED_LED);

        /* Turn on green LED to indicate burn
            complete */
        WRITEREGW(PIO_SODR, GREEN_LED);
    }
    /*
        Entry-point routine
    */
    int main(void)
    {
        /* Burn the user area of flash */
```

```
LoadFlash(flash_contents);

/* Verify the flash contents. If it didn't
    verify OK, then turn on the red LED. */
if (memcmp(flash_contents, (unsigned char *)
 FLASH_START, FLASH_SIZE))
    WRITEREGW(PIO_SODR, RED_LED);

/* Our work is done; halt. */
while (1);
}
```

The AT29LV1024 flash chip is programmed in 128-word sectors. For each sector write operation, the unlocking sequence is first sent to the chip (see the discussion on the EB40 hardware at the start of this chapter). The program then writes 128 words to the chip. This 128-word sector is temporarily stored in an internal RAM programming buffer within the flash chip. Each high-to-low transition of the write enable pin on the flash chip resets and starts the internal programming timer. If more than 150 microseconds goes by without the timer being reset, the auto-programmer is triggered, and it erases and reprograms the target sector. (If fewer than 128 words were loaded into the programming buffer, the uninitialized words will be erased to their default value of 0xFFFF.) While the auto-programmer is operating, reading any location in the device will show toggling values on the I/O_6 and I/O_{14} data lines of the device. Thus, your flash-loader code can poll for the end of the programming operation by simply reading any single flash location twice and comparing the results. If the two read operations yield the same result, then the programming operation is complete and you can proceed to the next sector.

You should note from the above explanation that the method the chip uses to signal that a programming operation is in process precludes you from running the flash-loader program out of the flash chip that is being burned. While not all flash memory devices have this restriction, most of them do—so if you intend your system to be self-reprogrammable you must generally ensure that it has enough RAM to entirely contain a flash-loader stub. For some mystifying reason, this issue frequently causes long threads of questions from novice programmers asking how they can circumvent the restriction and run the flash-loader directly out of the device being programmed. Presumably there are a terrifyingly large number of designs out there where someone built the hardware without considering that additional code RAM would be required in order to run a flash-loader. You can't work around this restriction, at least for the AT29LV1024 device and indeed the vast majority of flash devices in use today. If you need to run code out of flash and simultaneously write back to that same flash device (for instance, if you have a filesystem as well as directly executable code in the device), you need to look at multilayer flash devices. Sharp and Intel, among other manufacturers, have parts that satisfy this design need.

Our small program above erases and writes the entire user-accessible portion of the flash chip from the data included at compile time in boot.s, and then verifies the result. If the write and verify operations were successful, the green LED will come on at the end of this process. If the write completed, but the verify operation failed, the red and green LEDs will be illuminated. Any other combination of LEDs indicates that the code is stuck in the programming loop somewhere, or some other problem occurred.

By the way, this program is also important for another reason besides its basic functionality—it's the first example we have discussed so far that has explicitly used a function out of the standard libraries—memcmp().

The sourcecode/flash directory includes, along with the flash-loader sourcecode, an example flash.bin file for testing purposes; this file simply contains a text string. To verify the operation of the flash-loader, you should make, load and continue the program, exactly as in the previous examples. Once the green LED comes on to signal the end of the programming operation, press Ctrl-C to return to the gdb prompt. We'll now use gdb to peek around in memory and satisfy ourselves that the flash device was programmed successfully.

We'll use the x command (described in the gdb reference section earlier) to look at the user area of flash memory. Since I told you that the sample data we just loaded into flash memory was a text string, we should inspect it using the command x/s 0x01010000. This yields the following output:

```
0x1010000: "This is an example file to be
burned into flash memory.\n\n"
```

By comparison, we can also dump the same data to the debugging terminal in character format with the command `x/58c 0x01010000`:

```
0x1010000: 84 'T'  104 'h' 105 'i' 115 's' 32 ' '  105 'i' 115 's' 32 ' '
0x1010008: 97 'a'  110 'n' 32 ' '  101 'e' 120 'x' 97 'a'  109 'm' 112 'p'
0x1010010: 108 'l' 101 'e' 32 ' '  102 'f' 105 'i' 108 'l' 101 'e' 32 ' '
0x1010018: 116 't' 111 'o' 32 ' '  98 'b'  101 'e' 32 ' '  98 'b'  117 'u'
0x1010020: 114 'r' 110 'n' 101 'e' 100 'd' 32 ' '  105 'i' 110 'n' 116 't'
0x1010028: 111 'o' 32 ' '  102 'f' 108 'l' 97 'a'  115 's' 104 'h' 32 ' '
0x1010030: 109 'm' 101 'e' 109 'm' 111 'o' 114 'r' 121 'y' 46 '.'  10 '\n'
0x1010038: 10 '\n' 0 '\0'
```

You can experiment with the other formatting options to see what kind of output they produce.

While we're inspecting memory, let's also introduce another very useful command—disassemble *start-address end-address*. This command provides a symbolic disassembly of the specified memory region, pausing on page boundaries if necessary. For example, we can disassemble the vector area at the bottom of memory using the command disassemble 0 0x20, which will show something like this:

```
Dump of assembler code from 0x0 to 0x20:
0x0:    ldr    pc, [pc, #24]     ; 0x20
0x4:    ldr    pc, [pc, #24]     ; 0x24
0x8:    ldr    pc, [pc, #24]     ; 0x28
0xc:    ldr    pc, [pc, #24]     ; 0x2c
0x10:   ldr    pc, [pc, #24]     ; 0x30
0x14:   ldr    pc, [pc, #24]     ; 0x34
0x18:   ldr    pc, [pc, #-3872]  ; 0xffffff100
0x1c:   ldr    pc, [pc, #24]     ; 0x3c
End of assembler dump.
```

As you can see, Angel initialized all the entries in the vector area to use a secondary jump table immediately after the real table. The only exception is the IRQ vector, which is directed via an Atmel-specific interrupt controller register named AIC_IVR. For more information on how this interrupt mechanism works, consult the AT91R40807 datasheet's description of the Standard

Interrupt Sequence (this information is found on p.71 of the current version at the time of writing).

One excellent reason for using a secondary vector table is so that the interrupt handler code can be at any arbitrary location in memory, without needing to be within the boundaries of a single-word relative branch instruction. A second possible reason is that it's easier to hook interrupts in realtime if you can simply write a pointer rather than having to synthesize a machine code instruction. A third good reason for using this technique is that it's much easier simply to write an absolute function pointer than to try and calculate a relative offset and synthesize a branch instruction. Let's just inspect the secondary table to satisfy our curiosity as to where those exceptions are being vectored, with the command x/8xw 0x20:

```
0x20:    0x02073748   0x020735e4   0x02073768   0x02073754
0x30:    0x02073758   0x0207375c   0x02068268   0x02073764
```

As you can see, these vectors are all in the upper half of the EB40's SRAM, inside the area reserved for Angel's use.

A Simple ROM-Startup Program

Now that we have the technology to burn our programs into ROM, let's develop our first self-sufficient application and load it onto our board. The first thing we need to do is write a new linker script that directs the code and data appropriately. A suitable linker script is provided below:

```
/*

Linker script for Atmel EB40

This is an introductory linker script for
programs that use ROM-based startup

LewinA.R.W. Edwards (sysadm@zws.com), Feb-2002

*/
/*

The entry-point is the power-on reset
vector at the start of the vector table.

*/
```

```
ENTRY(vectors)

SEARCH_DIR(.)

/*
    This simple partial example only shows the
    boot configuration of the device, without
    remapping. It also does not show the off-
    chip SRAM (since this is inaccessible
    before remapping).
*/

MEMORY
{
    /* 64KBytes of user flash */
    flash   : org = 0x00000000, len = 0x00010000

    /* 128K on-chip SRAM */
    sram128k : org = 0x00100000, len = 0x00020000

    /* 8K on-chip SRAM */
    sram8k : org = 0x00300000, len = 0x00002000
}

SECTIONS
{
    .text :
    {
        *(.text);
        . = ALIGN(4);
        *(.glue_7t);
        . = ALIGN(4);
        *(.glue_7);
        . = ALIGN(4);
        etext  =  .;
    } > flash

    .data :

    {
        datastart = .;
        __data_start__ = . ;
        *(.data)
```

```
        . = ALIGN(4);
        __data_end__ = . ;
        edata  =  .;
        _edata  =  .;
} > sram128k

.bss :
{
        __bss_start__ = . ;
        *(.bss); *(COMMON)
        __bss_end__ = . ;
} > sram128k
}
```

This linker script is not a complete general solution for writing ROM-startup applications, but it suffices for simple assembly-language programs like ours, and we will improve on it later.

Since we'll be loading this program onto the board via the simple flash-loader described in the previous section, we need to convert the ELF output we've been using in all the previous examples into a raw binary format suitable for burning into flash memory. We accomplish this by adding an arm-elf-objcopy command to the makefile. The final makefile is shown below:

```
# Embedded Systems Development on a Shoestring
# Example ROM startup program
#
# Lewin A.R.W. Edwards (sysadm@zws.com), Jun-2002.

ASFLAGS = -mcpu=arm7tdmi -gstabs
LDFLAGS = -Teb40-rom.ld -nostartfiles -Lgcc -L.
OBJS    = boot.o
EXE     = blink_rom.elf

$(EXE): $(OBJS)
        arm-elf-gcc $(LDFLAGS) -o $(EXE) $(OBJS)
        arm-elf-objcopy -O binary $(EXE) flash.bin

%.o:%.s
        arm-elf-as $(ASFLAGS) $< -o $@

clean:
        rm -f $(OBJS)
```

```
rm -f $(EXE)
rm -f flash.bin
```

Now, you *could* simply use the boot.s file from the original RAM-based LED blinking application, compile and link it using the above makefile and linker script, and the resulting binary will run directly out of flash without modification. However, you wouldn't be able to tell it's doing anything, because it won't actually blink the LED. This is because the RAM-based program ran in an environment that had been set up by the Atmel bootloader and Angel; among other things, the bootloader mapped GPIO ports P1, P2 and P4 (the red, green and yellow LEDs, respectively) to the control of the PIO Controller, and set these port bits into output mode. In order for our program to yield the expected results, we need to add a little more power-on initialization code to boot.s in order to set up the environment:

```
@@@@@@@@@@@@@@@@@@@@@@@@@@@@@@@@@@@@@@@@@@@@@@@@@@@@@@@@@@@@
@ Embedded Systems Development on a Shoestring
@
@ EB40 LED Flasher (ROM-startup version)
@
@ Lewin A.R.W. Edwards (sysadm@zws.com), Jun-2002.
.section .text
.code 32
.globl vectors

@@@@@@@@@@@@@@@@@@@@@@@@@@@@@@@@@@@@@@@@@@@@@@@@@@@@@@@@@@@@
@ This is the ARM interrupt vector table. Our program
@ only implements the power-on reset vector; any other
@ exception will hang the system.
vectors:
        b reset    @ Reset
        b .        @ Undefined instruction
        b .        @ SWI
        b .        @ Prefetch abort
        b .        @ Data abort
        b .        @ reserved vector
        b .        @ irqs
        b .        @ fast irqs
```

```
@@@@@@@@@@@@@@@@@@@@@@@@@@@@@@@@@@@@@@@@@@@@@@@@@@@@@@@@@@
@ Entry-point code
reset:
     ldr r6,=0x02
     @ Bit 1 is the red LED on the EB40

     @ Direct red LED output bit (P1) for PIO controller
     ldr r4,PIO_PER
     str r6,[r4]

     @ Enable output on GPIO bit P1
     ldr r4,PIO_OER
     str r6,[r4]

     ldr r4,PIO_SODR
     ldr r5,PIO_CODR

blink_loop:
     str r6,[r4]                    @ Turn on LED

     @ Brief pause
     ldr r0,delay_constant
     ldr r1,=0
pause_loop1:
     sub r0,r0,#1
     cmp r0,r1
     bne pause_loop1

     str r6,[r5]      @ Turn off LED

     @ Brief pause
     ldr r0,delay_constant
pause_loop2:
     sub r0,r0,#1
     cmp r0,r1
     bne pause_loop2

     @ Loop forever
     b blink_loop

@@@@@@@@@@@@@@@@@@@@@@@@@@@@@@@@@@@@@@@@@@@@@@@@@@@@@@@@@@
@ Miscellaneous constants
PIO_PER:      .word 0xffff0000
              @ PIO Enable Register
```

```
PIO_OER:        .word 0xffff0010
                @ PIO Output Enable Register
PIO_SODR:       .word 0xffff0030
                @ Set Output Data Register
PIO_CODR:       .word 0xffff0034
                @ Clear Output Data Register
delay_constant:  .word 0x00040000
                @ Provides a decent delay period
```

Once you've built this project with make, you need to combine it with the flash-loader from the previous section in order to get it onto the board. If you haven't already done so, copy the sourcecode/flash directory to your hard drive. Copy the flash.bin file you just generated above into this directory, overwriting the demonstration flash.bin file.

Now change to the flash-loader directory, run `make clean ; make all` and use `arm-elf-gdb flash.elf` followed by the usual `set remotebaud 9600, target rdi / dev/ttyS0`, and `load` commands to load your custom flash-loader code into the EB40's RAM. Finally, use the `continue` command to start the burn operation. Wait for the red LED to go out, and your board is ready to boot!

To run the code you just flashed onto the board, move switch SW1 to the UPPER MEM position and press the reset button or cycle power; the LED should start blinking immediately. Notice, by the way, that the blink rate is approximately 0.5Hz; almost exactly a quarter the speed of the exact same code running out of RAM. This is because of wait states set on the flash chip slowing down the processor. In order to get the absolute best performance out of your code, you should copy it to the internal 32-bit, zero-wait-state SRAM and run it from there.

Before proceeding, make sure you move SW1 back to the LOWER MEM position and press the reset button, so that Angel is back in control.

A Complete ROM-Startup Application in C

The ROM-based LED blinking program above demonstrates the bare minimum infrastructure you need in order to get a simple assembly-language program loaded into and started up out of on-board ROM. However, it lacks a lot of functionality that we'll require for developing more complex applications. The example we're going to build in this section is almost a complete framework for a fully-featured ROM-based application written in C. (There are a few other things you would probably want to do in a real application, such as setting up the flash memory wait states to make more efficient use of memory bandwidth and so on, but these are mostly cosmetic details.)

The first step in achieving this goal is to modify the linker script somewhat:

```
/*
    Linker script for Atmel EB40 ROM-startup
        applications
    Lewin A.R.W. Edwards (sysadm@zws.com),
        Feb-2002
*/
ENTRY(vectors)

SEARCH_DIR(.)

SECTIONS
{
    .text 0x00000000 :
    {
        *(.text);
        . = ALIGN(4);
        *(.glue_7t);
        . = ALIGN(4);

        *(.glue_7);
        . = ALIGN(4);
        *(.rodata);
        . = ALIGN(4);
        etext  =  .;
    }
    .data 0x00300000 : AT (ADDR(.text) +
    SIZEOF(.text))
```

```
{
        . = ALIGN(4);
        datastart = .;
        __data_start__ = . ;
        *(.data)
        . = ALIGN(4);
        __data_end__ = . ;
        edata  =  .;
        _edata  =  .;
}

.bss 0x00300000 + SIZEOF(.data) :
{
        __bss_start__ = . ;
        *(.bss); *(COMMON)
        __bss_end__ = . ;
        . = ALIGN(4);
        . += 0x800 ;
        _stack_top = . ;
}

__data_rom_start__ = LOADADDR(.data) ;
}
```

There are two major changes you'll see in this script compared to the other scripts we've examined so far. By far the more important change you will observe is that the .data section is relocated to a VMA in the 8K of on-chip SRAM, even though its information is still emitted into the executable at an LMA immediately after the .text section.

Note that since the .data segment is being emitted to a different LMA than VMA, we now need to define three symbols to keep track of it. The symbols __data_start__ and __data_end__ define the starting and ending VMA, respectively, of the initialized data segment (and by inference, its size). The symbol __data_rom_start__ points to the physical load address (LMA) of the information that needs to be copied to the .data section's VMA.

The second big change in the linker script is a largely cosmetic difference—instead of using named memory regions, I have written the above script to work entirely with absolute addresses.

This is a personal preference; in situations where the VMA and LMA are not the same, I find it improves clarity for me to work directly with absolute memory addresses. If you feel more comfortable using the *>vma-region* AT*>lma-region* syntax on the output section directive, then by all means do so; there's no major functional difference between the two methods.

```
@@@@@@@@@@@@@@@@@@@@@@@@@@@@@@@@@@@@@@@@@@@@@@@@@@@@@@@@@@@
@ Embedded Systems Development on a Shoestring
@
@ EB40 General-Purpose ROM Application Startup
@
@ Lewin A.R.W. Edwards (sysadm@zws.com), Jun-2002.
.section .text
.code 32
.globl vectors

@@@@@@@@@@@@@@@@@@@@@@@@@@@@@@@@@@@@@@@@@@@@@@@@@@@@@@@@@@@
@ This is the ARM interrupt vector table. Our program
@ only implements the power-on reset vector; any other
@ exception will hang the system.
vectors:
        b reset           @ Reset
        b exception       @ Undefined instruction
        b exception       @ SWI
        b exception       @ Prefetch abort
        b exception       @ Data abort
        b exception       @ reserved vector
        b exception       @ irqs
        b exception       @ fast irqs

@@@@@@@@@@@@@@@@@@@@@@@@@@@@@@@@@@@@@@@@@@@@@@@@@@@@@@@@@@@
@ Exception error
exception:
        ldr r4,PIO_SODR
        ldr r5,PIO_CODR

        ldr r6,=0x16 @ All three LEDs
blink_loop:
        str r6,[r4]  @ Turn on LEDs
```

```
                @ Brief pause
                ldr r0,delay_constant
                ldr r1,=0
pause_loop1:
                sub r0,r0,#1
                cmp r0,r1
                bne pause_loop1

                str r6,[r5]  @ Turn off LED

                @ Brief pause
                ldr r0,delay_constant
pause_loop2:
                sub r0,r0,#1
                cmp r0,r1
                bne pause_loop2

                @ Loop forever
                b blink_loop

@@@@@@@@@@@@@@@@@@@@@@@@@@@@@@@@@@@@@@@@@@@@@@@@@@@@@@@@@@@@@@@
@ Entry-point code
reset:
                @ First, set up the PIO controller to
                map P1,2,4,5
                ldr r6,=0x36  @ Our desired GPIO bits

                @ Map relevant bits to GPIO controller
                ldr r4,PIO_PER
                str r6,[r4]

                @ Enable output on GPIO bits associated
                   with LEDs
                ldr r6,=0x16
                ldr r4,PIO_OER
                str r6,[r4]

                ldr r5,PIO_SODR
                ldr r7,PIO_CODR

                @ Turn off all LEDs
                str r6,[r7]
```

```
        @ Turn on red LED
        ldr r6,=0x02
        str r6,[r5]

        @ Copy the initialized data section
          from LMA to VMA
        ldr r1, data_start
        ldr r2, data_end
        ldr r3, data_source
initialize_data:
        cmp r1, r2
        bge init_data_done
        ldrb r4, [r3],#+1
        strb r4, [r1],#+1

        b initialize_data
init_data_done:
        @ Turn off red LED
        str r6,[r7]

        @ Turn on yellow LED
        ldr r6,=0x10
        str r6,[r5]

        @ Zero out the .bss section
        ldr r1, bss_start
        ldr r2, bss_end
        ldr r3, =0
clear_bss:
        cmp r1, r2
        strne r3, [r1],#+4
        bne clear_bss

        @ Turn off yellow LED
        str r6,[r7]

        @ Turn on green LED
        ldr r6,=0x04
        str r6,[r5]
```

```
init_main:
        @ Initialize the stack pointer
        ldr r13,stack_pointer

        @ Call main function
        bl main

        @ Hang if main() exits
        b .

@@@@@@@@@@@@@@@@@@@@@@@@@@@@@@@@@@@@@@@@@@@@@@@@@@@@@@@@@@@@@@@
@ Miscellaneous constants
delay_constant:   .word 0x20000
stack_pointer:    .word _stack_top
bss_start:        .word __bss_start__
bss_end:          .word __bss_end__
data_start:       .word __data_start__
data_end:         .word __data_end__
data_source:      .word __data_rom_start__
PIO_PER:          .word 0xffff0000
                  @ PIO Enable Register
PIO_OER:          .word 0xffff0010
                  @ PIO Output Enable Register
PIO_SODR:         .word 0xffff0030
                  @ Set Output Data Register
PIO_CODR:         .word 0xffff0034
                  @ Clear Output Data Register
.end
```

As you can see, this startup code builds on the earlier example principally in that it copies the data section across from ROM to RAM before passing control to the main C program. This means, among other things, that this piece of code is "pure"[24] —it can be restarted without needing to reload the entire program.

Another feature that has been added to this code is a little subroutine to inform you if the processor hits an exception. Any exception will cause all three LEDs to flash on and off forever,

---

[24] This isn't *necessarily* correct for all programs written like this (since the hardware state won't necessarily be the same on a subsequent run unless there is a hardware reset), but it happens to be true for our simple program.

rather than simply hanging the board silently as in previous examples.

In order to demonstrate that the .data section is in fact being initialized correctly, I've written a simple C program to go along with this startup code. This program is listed below.

```
/*

    Embedded Systems Development on a Shoestring
    A more complex ROM-based program
    Lewin A.R.W. Edwards (sysadm@zws.com), Jun-2002.

*/

/* GPIO control registers */
#define PIO_PER      0xffff0000
#define PIO_OER      0xffff0010
#define PIO_SODR     0xffff0030
#define PIO_CODR     0xffff0034
#define PIO_PDSR     0xffff003c

/* Bits to be set/reset on the GPIO port in
   order to toggle the LEDs */
#define RED_LED      0x00000002
#define YELLOW_LED   0x00000010
#define GREEN_LED    0x00000004

/* Input bit definitions */
#define BUTTON_SW4 0x00000020

/* Quick and dirty macros to write a register or
   memory location word, halfword or byte */

#define WRITEREGW(addr,value) *((volatile
unsigned int *)
(addr)) = (value)
#define WRITEREGH(addr,value) *((volatile
unsigned short *)
(addr)) = (value)
#define WRITEREGB(addr,value) *((volatile
unsigned char *)
(addr)) = (value)

/* Quick and dirty macros to read a register or
   memory location word, halfword or byte */
```

```
#define READREGW(addr) (*((volatile unsigned int *)
(addr)))
#define READREGH(addr) (*((volatile unsigned short *)
(addr)))
#define READREGB(addr) (*((volatile unsigned char *)
(addr)))
/* This variable identifies the current LED
   being displayed */
int lednum = 2;
/*
    Main function
*/
int main(void)
{
    while (1) {
        /* Turn on one LED corresponding to the
           lednum variable */
        switch (lednum) {
            case 0 :
            WRITEREGW(PIO_CODR, YELLOW_LED |
             GREEN_LED);
            WRITEREGW(PIO_SODR, RED_LED);
            break;

            case 1 :
            WRITEREGW(PIO_CODR, RED_LED |
             GREEN_LED);
            WRITEREGW(PIO_SODR, YELLOW_LED);
            break;

            case 2 :
            WRITEREGW(PIO_CODR, RED_LED |
             YELLOW_LED);
            WRITEREGW(PIO_SODR, GREEN_LED);
            break;

            default:
            WRITEREGW(PIO_CODR, RED_LED |
            YELLOW_LED | GREEN_LED);
            break;
        }
```

```
/* Wait for user to press and release SW4 */
while (READREGW(PIO_PDSR) & BUTTON_SW4) { }
while (!(READREGW(PIO_PDSR) & BUTTON_SW4)) { }
lednum++;
if (lednum > 2)
    lednum = 0;
}

}
```

You can load this example onto your EB40 using the same method as I described for the previous ROM-startup program; compile this program to generate a flash.bin file, merge it with the flash-loader project, and run the result in order to write the new code to flash memory. Flip SW1 to UPPER MEM and reset the board to run this program.

This example illuminates one of the LEDs; pressing the SW4 button cycles the currently illuminated LED from red to yellow to green, then back to red. By default, the program starts with the green LED illuminated. To satisfy yourself that everything is working as it should, you can change the line int lednum=2; to int lednum=1; then recompile and reload the code. The default LED on powerup or hardware reset will now be yellow.

Blind-Debugging Your Program

Let's digress for a moment and think in detail about debugging. Most of your debugging will be performed "blind" because it will be running on the real hardware, probably without debugger support. For example, all the ROM-startup programs mentioned in this book have to be debugged without the help of gdb. You'll see that in the preceding firmware example, I indicate progress through the startup code by cycling LEDs on the board (the initialization operations are so fast that you can't see this happening). In fact, the reason I added this code is because I initially struck a seemingly strange problem—I wrote the entire program (with no progress indicator code) assuming it to be trivial, only to find that it didn't work at all. I didn't want to break out a hardware debugger for such a simple program, so I added the progress indicator and found that the program was halting sometime after setting the red LED on, but before getting as far as turning on the yellow LED. It's at this point that I added the code that shows you if the processor encounters an exception. This code revealed that the reason the program was halting was indeed that it was hitting an exception.

Further testing (basically just moving a "b ." instruction down the program and recompiling) narrowed the crash down to the `strb r4, [r1],#+1` instruction. On close inspection, it transpired that I had a typographical error in the linker script. This error was causing my program to write to an unimplemented memory area, thereby causing a data abort exception.

These sorts of manual debugging techniques were used very commonly in the days of older 8-bit and even 16-bit home computers, where the hardware had little or no special debugging support and it was almost universal practice to develop a program in vivo within the actual system that was intended to run the final version. These methods have an even longer and more honorable history on earlier platforms, such as minicomputers, that had handy front panel lamps to assist with the task. Unfortunately, these skills seem to have fallen by the wayside, at least from the perspective of software engineers; when I discuss this debugging style with people who work exclusively in software, they express puzzlement. The modern software engineer is apparently trained to rely on symbolic debuggers. The reason the loss of the older debugging art is so regrettable is that it's a lowest common denominator method that is practically always available, no matter what your budget for development hardware or the idiosyncrasies of your target platform. In my experience, engineers with a hardware bent understand and take to this method of debugging easily. This may simply be because these engineers have more and longer familiarity with working on systems where much of the state cannot directly be examined, but must be inferred by inspecting externally visible signals that are affected by the items of interest and working backwards to an understanding of the machine's internal state. Whatever the underlying reasons, the skill of debugging through the smallest of viewports is absolutely essential to the funds-constrained developer who cannot afford exotic hardware-based debugging tools.

Thus, if you don't have a JTAG ICE debugging module, or the software for your debugger module doesn't support symbolic debugging of the executables generated by the GNU toolchain, you have a few options. The most powerful process is to port a debugger back-end to your hardware, so that you can include it in the device's real firmware right up until the final build. Gdb stubs have been ported to the Atmel EB40 already; you can find information on how to install them as part of the Red Hat RedBoot documentation.

For some targets, however, it will be impractical to do this, and you have to rely on implicit debugging techniques like that described above. There's a limit to exactly how much you can debug with just a few LEDs to let you know your program's status—at least, if you want to keep your sanity and ship your code inside a reasonable timeframe. For this reason, you'll want to find a more flexible channel for communicating program status information. The method I prefer to use is to implement a serial port (even if the final device won't have one) and use that to dump trace information, memory contents and so forth manually. This is rather less convenient than using a debugger, since we can only change what we're looking at by recompiling and reloading the program, but it is still an extremely powerful tool.

In the sourcecode/serial_demo directory on the CD-ROM, you'll find a small example collection of routines to dump memory and trace information on a serial console attached to the second serial port (SERIAL B) of the EB40. Note that this port is wired differently from SERIAL A; you'll have to connect this to your PC via a nullmodem cable, not a straight-through cable. You can modify this program to output on SERIAL A by editing serial.h and changing the USART_BASE macro to 0xFFFD0000.

This particular example is built for RAM startup, so that you can load and test it easily and quickly without bothering to burn it into flash. Although it's not a comprehensive library of every conceivable routine you will need, it is illustrative of the kind of routines you'll find useful. A module like this is one of the first things I try to get running on any new hardware platform, because it simplifies debugging enormously.

Below is a listing of these serial routines. The only restriction on these routines is that you must call SER_Init(*baudrate*) before using any of the other code.

```c
/*

     Embedded Systems Development on a Shoestring
     Serial port output code for Atmel EB40
     Lewin A.R.W. Edwards (sysadm@zws.com), Jun-2002.
*/

#include <stdlib.h>

#include "serial.h"

/*

     Initialize USART to desired baud rate
*/
void SER_Init(int baudrate)
{
     /* Disable USART interrupts */
     WRITEREGW(USART_IDR, 0x000003ff);

     /* Reset the USART and enable transmitter and
        receiver */

     WRITEREGW(USART_CR, 0x0000015c);

     /*  Set the mode register. This sets the USART
         to asynchronous mode, 8N1, clock source
         MCK, normal mode */
     WRITEREGW(USART_MR, 0x000008c0);

     /* Set the baud rate generator */
     WRITEREGW(USART_BRGR, (32768000 / (16 *
       baudrate)));
}
/*

     Write character to serial port
*/
void SER_WriteChar(char thechar)
{
     /* Wait for the transmit holding buffer to
        be empty */
```

```c
    while (!(READREGW(USART_CSR) & 0x00000002)) {}

    /* Send character to the host */
    WRITEREGW(USART_THR, (int) thechar);
}
/*
    Write an ASCIIZ string to the serial port
*/
void SER_WriteString(char *thestring)
{
    char *p = thestring;

    while(*p)
        SER_WriteChar(*(p++));
}
/*
    Dump a section of memory to the serial port
*/
void SER_DumpMemory(char *memory, int length)
{
    int i,j;
    char tmps[16];
    for (i=0;i<length;i+=16) {
        sprintf(tmps,"%-08.81X: ", ((int)
         memory) + i);
        SER_WriteString(tmps);
        for (j=0;j<16 && (j+i < length);j++) {
            sprintf(tmps,"%-02.2hX",
             memory[i+j]);
            SER_WriteString(tmps);
            if (j<15 && (j+i < length - 1))
                SER_WriteString(", ");
        }
        SER_WriteString("\n\r");
    }
}
```

(By the way, if you look at the linker script you'll notice one new thing—an input section definition that loads .rodata* sections into the .text output section. This is because constant strings declared in gcc are emitted into .rodata sections.)

Miscellaneous Glue—Handling Hardware Exceptions in C with gcc

You may have had some experience with another embedded C compiler that provides an "interrupt" keyword and some kind of API that allows you to vector a hardware exception directly to your C code. For example, most compilers that target MS-DOS support an interrupt keyword to qualify function prototypes, and they allow you to use either DOS int 21h services or direct vector read/write operations to vector hardware and software exceptions to your own code.

However, gcc lacks this sort of feature by design. The stated rationale for this is that the main reason for writing in a high-level language is in order to generate portable code; anything that relies on some specific type of hardware exception being available is inherently non-portable and therefore there is no place in a high-level language for such a low-level construct.

Working around this deficiency is extremely easy. We simply write a tiny assembly-language stub that saves the processor context on the stack and calls a C function to handle the actual interrupt condition. Here's a small piece of example code to achieve this on an ARM platform:

```
fiq_handler:
        stmdb r13!, {r0-r7, r12}
        ldr r12,irqvector
        bx r12
        ldmia r13!, {r0-r7, r12}

        @ return to interrupted code
        subs pc, r14, #4

fiqvector:
        .word c_fiq_handler
```

To use this code, simply vector the desired exception to fiq_handler, and provide an external C function named c_fiq_handler that performs the actual interrupt handling. Observe that we do not need to save the entire processor register set, because the ARM architecture contains a set of banked (shadow) registers r8-13 dedicated to use in FIQ mode. For more information on this topic, you should refer to the ARM7TDMI core datasheet I mentioned earlier.

There's slightly more to this story than the simple explanation above, however. To begin with, in all the programs we've illustrated so far we have only set up a single stack. The ARM processor actually maintains four stacks; one for normal user mode, one for SVC mode, one for IRQ mode, and one for FIQ mode. If we're going to allow our processor to handle exceptions, we have to set up some or all of these stacks. This is something you would normally do in the assembly-language startup stub, before initializing the main program. An example code snippet is presented below:

```
@ Put CPU into IRQ mode and set up IRQ stack
mrs  r0,cpsr
bic  r0,r0,#0x1f
orr  r0,r0,#0x12
msr  cpsr,r0                    @_cf
ldr  r13,irqstack

@ put CPU into SVC mode and enable the IRQ
@ interrupt
mrs  r0,cpsr
bic  r0,r0,#0x9f
orr  r0,r0,#0x13
msr  cpsr,r0                    @_cf

@ setup SVC stack
ldr  r13,svcstack

@ put CPU into FIQ mode and set up FIQ
@ registers and stack
ldr  r0,const_d1
msr  cpsr,r0                    @_cf
ldr  r8,=0
ldr  r9,=0
ldr  r10,=0
ldr  r11,=0
ldr  r12,=0
ldr  r13,fiqstack
```

```
     [ ... remainder of initialization code ... ]
     @ Jump to main C program
     bl main
fiqstack:  .word _fiq_stack_top
irqstack:  .word _irq_stack_top
svcstack:  .word _svc_stack_top
```

(You would then define the locations _fiq_stack_top, _irq_stack_top and _svc_stack top in your linker script).

Note that the ARM core only directly supports two external interrupt sources—IRQ and FIQ. Almost all ARM implementations, therefore, have an off-core interrupt controller that extends this simple model to support prioritized interrupts from various on-chip peripherals. The AT91 series' interrupt controller is quite powerful, and allows you to have up to 32 separate interrupt vectors. Unlike many other implementations, the Atmel interrupt controller even handles the vectoring process for you. (Usually on ARM platforms, you vector the IRQ handler to a piece of code that inspects a register in the interrupt controller to determine what is the highest-priority pending interrupt. This code then services the interrupt, clears the pending flag, and inspects the interrupt register again to see if a lower-priority interrupt is pending. This process is repeated until all pending interrupts have been serviced). For more information on the interrupt controller module in the AT91R40807, refer to the product datasheet. Using this peripheral is extremely simple by comparison with the interrupt controllers found in many other 32-bit parts, and overall it's a most impressive piece of design.

## Chapter 5

Portability and Reliability Considerations

Throughout this text, I have frequently made mention of temporary solutions, such as prototyping around an embedded PC with the intention of porting your code to real hardware later (e.g., after securing investment capital). Approaches like this can save you a lot of time in creating your first working prototype, and in many cases there is absolutely no other way that the small one-person engineering "team" will be able to bring a project to completion in time to generate enough income to stay in business.

Your task of migrating from demonstration hardware to the real circuit will be facilitated greatly if you keep the migration goal in mind at all times while designing the firmware. Some of the factors you should consider are described below. Mostly, these are simply good design practices to follow in embedded systems development, but they bear spelling out here because the initial version of your application will be running in a (relatively) unconstrained environment. On the demonstration hardware, the temptation will be strong to cut corners and to exploit the lax restrictions of the platform; you will have ample RAM, fast processors, possibly a hard disk for local storage of temporary files, probably a multitasking environment, and so on. The less discipline you use here, the more time and effort it will take to port your code to the final real hardware, and the harder it will be for you to be able to guarantee your code's reliability in its new environment. People working on tiny 8-bit embedded platforms are used to rigorous design and memory tracking; people working on 32-bit embedded platforms (especially if they come from a nonembedded programming background) aren't accustomed to these issues, but they are just as important on these larger platforms.

Anecdotes about this sort of issue are tragically common in the embedded world, but I will give you a single short story here to illustrate the evils of incomplete planning. In 2001, I was asked to build a special one-off demonstration version of a complicated multimedia appliance. This was a rush job, and the only stated goal was to provide a couple of demonstration pieces to ship to a customer. The theory was that we would get an order from the customer and use it to fund development of a real product. (The customer was informed about this process, of course.) The prototypes were built around an off-the-shelf Intel SBC running Linux, and the code was pulled together with the bare minimum of functionality, because it was stated from the first day of this project that all the old code would need to be replaced if the device went into production. (We intended to build our own hardware platform around a different CPU on a custom board, also.)

Everything would have been fine if we had only ever shipped those two prototypes, but other customers found out about this product, and since it had a healthy profit margin, we sold a few hundred units. So, although it was clearly understood at all times that the existing software was a hack and far from ideal, once the product "escaped" our doors in this way, it immediately began to generate a snowball of unavoidable maintenance work. Since we couldn't leave the existing customers waiting for six months for the real product (and real firmware) to become available, we had to divert development resources towards maintaining the old codebase, fixing bugs, working around undesirable behaviors, and adding new features at customer request. It was almost a year before we reached a point with the old codebase where we could realistically tell all existing customers that they should live with their current firmware version, and that there would be no more running updates until the all-new "real" firmware became available.

The original prototype took approximately two months to build. The new code took about three and a half months to get to a point equivalent to the old prototype, plus about one and a half months' engineering time on the hardware. Even allowing for the fact that we were more familiar with the hardware platform on this second attempt, we would probably have only been two or three months later in delivering the original prototypes if we had done them "properly"—i.e., with forward-looking, maintainable code.

Realistically, there are occasions when that extra couple of months' development time is going to mean the difference between being able to maintain a positive cash flow or going out of business, and in fact the story I just related was close to being such a scenario. However, hopefully tales like the one above will encourage you to spend as much effort as possible at the outset of a project to avoid wasted work that will need to be repeated with more rigor later. The rules below are a good starting point.

Avoid dynamic memory allocation wherever possible. Your evaluation platform probably has significantly more RAM than the final product will boast. It may well also have a demand-paging virtual memory manager that makes it appear to have practically unlimited available memory. However, no matter how intelligent your OS's memory manager might be, random allocation and deallocation of memory chunks inevitably results in memory fragmentation[25]. You need to have a reasonable idea of how much RAM the final system is going to require anyway (in order to build the hardware!) so, as far as possible, work it out beforehand and allocate buffers, stacks and other structures statically.

Some types of variables don't lend themselves directly to compile-time static memory allocation. For instance, many applications need to maintain arbitrarily-sized, dynamically resizable arrays of structures, each of which describes a file, event, network connection or other logical entity. The number of entries required in these arrays is impossible to predict at compile-time and is expected to vary during a single run of the program. One common way of implementing such structures on a "real" desktop system would be with a linked list, but this relies on a dynamic memory allocation manager. The technique of allocating and deallocating tiny chunks of memory like this practically guarantees that the application will eventually frag-

---

[25] There are some last-ditch ways of dealing with this, such as always referring to memory through double-indirected pointers a la MacOS (thereby allowing the memory manager to shuffle allocated blocks around behind your back), but these techniques are excruciating, time-wasteful and don't address the underlying logical problem adequately.

ment all available memory and will need to be shut down and restarted. This limitation appears to be quite reasonable to most programmers who write off-the-shelf application software, but is totally unacceptable in a high-availability embedded system.

One possible solution to this is simply to declare a flat array of structures, e.g. `struct mystruct[MAX_NUMBER_RECORDS]` and populate them as necessary. Although computer science lecturers would deplore it, this is an adequate solution when the records in your data structure will always be a fixed size. In many cases, however, the data records will be variable in length, and if you're storing them in a flat array, you need to size each element for the largest possible item to be contained. In such cases, it's much more efficient to reserve a simple memory array of whatever size you can spare, and pack the data elements as close as possible together in this space, with a special marker value to indicate where the list ends. It does mean slightly more effort, since you have to write functions to insert, remove and query from this packed list, but those functions are simple to write and quite fast.

Consider a list of fully qualified filenames that refer to an MS-DOS compatible filesystem (on a floppy disk, for the sake of argument) where the maximum possible path length is 128 characters, with the usual 8.3 character limit on filenames. For convenience, you would probably want to store these in memory in ASCIIZ format, so that means that a worst-case path length is 129 bytes. Suppose you've decided to allocate 8Kbytes of space for your filename list. If we follow the flat array method, we will create an array like this, which only gives us 63 spaces for pathnames:

```
#define MAX_FILENAME_LENGTH 129
#define FILENAME_SPACE 8192

struct
{
    char filename[MAX_FILENAME_LENGTH];
} file_list[FILENAME_SPACE / MAX_FILENAME_LENGTH];
```

However, stop and consider that few or no filenames are likely to reach the 128-character limit. Since users aren't likely to create more than one or two levels of nested directories on a floppy disk, and each directory and filename is inherently limited to twelve characters, most of the file paths we'll be working with are likely to be between thirteen to forty characters long. What a waste the underpopulated array is in this case! It's vastly more efficient to store the filenames in memory one after the other in a simple array of characters, and indicate the end of the list with a double NULL byte:

```
#define FILENAME_SPACE 8192

char file_list [FILENAME_SPACE];
```

Depending on the individual filename lengths, we could be able to store more than 2,000 entries using this method, with no additional RAM requirement. All we need to do is implement a few simple functions, such as the example below, which fetches a string from the structure, given a pointer to the start of the structure and an index number.

```
/*
    Retrieve entry #stringnum from ASCIIZ string
    list. Returns pointer to desired string, or
    empty string if there are fewer than
    stringnum items in the list.
*/
char *GetString(char *list, int stringnum)
{
    while (stringnum && *list) {
        while (*(list++));
        list++;
        stringnum—;
    }
    return list;
}
```

Similar techniques can be used for more complicated data structures. (Purely as an aside, the code above is very useful for internationalizing your product. If you store all the strings used by your user interface in a structure of the type illustrated above, you can easily change your display into another language by changing a single global pointer reference. You can also use external translation personnel to create foreign-language versions of your application without needing to reveal any sourcecode to these outside parties. Simply send them the text resource for translation, along with any special rules about maximum string length or spacing that your user interface requires.)

In other instances, you may have a piece of code that will allocate memory in different ways according to the input data it receives. You should be able to determine a "high water mark" indicating the *total* amount of memory a worst-case set of inputs will require (essentially by adding up all the malloc() calls made during such a worst-case run), but simply allocating a worst-case size for each of the possible buffers in the code snippet would require much more RAM than the "high water mark". For example, your code might require three different memory buffers to read in special parameters. The maximum size of any one of these parameters might be 4K, so a per-item-total worst-case requirement would be 12K. However, you know because of the nature of the function being implemented that it is impossible for all three of these buffers to be maximally allocated. Perhaps if buffer one receives a 4K structure, you can be sure that buffer three will only require a few bytes. You need a system that will give you the flexibility of being able to prevent collisions between numerous separate memory areas, without the long-term stability dangers inherent in a systemwide dynamic memory manager.

In the same vein, embedded systems quite commonly have sections of code that you know will never be running concurrently. Clearly, in a resource-constrained environment it is desirable to allocate enough memory only for the code path that requires the most space, and not reserve separate storage for two modules that will never need to share the system. It's much more efficient to allocate a single buffer and give each code segment exclusive access to that area when it's in control.

Both these latter issues can be addressed by implementing one or more private memory managers. Below is a simple example of such a memory manager, implemented in C.

```
#define DMEM_HEAP_SIZE    131072

/*

    These variables store the current state of the
    memory manager.
```

```
*/
unsigned char *heapstart;
unsigned char *heapend;
unsigned char *heapnext;

/*
    This is the actual heap.
*/
unsigned char heap[DMEM_HEAP_SIZE];

/*

    Power-on initialization for the memory manager.
    This should only be called once.
*/
void mmgr_Initialize(void)
{
    heapstart = heap;
    heapnext = heapstart;
    heapend = heapstart + sizeof(heap);
}

/*

    Allocate some "permanent" storage space.
    Memory allocated this way is allocated at
    the bottom of the heap and the heap start
    point is adjusted so that this memory will
    never be returned to the free pool. Re
    turns pointer to memory area or NULL if
    there is insufficient space.
*/
void *mmgr_PermAlloc(size_t allocsize)
{
    void *result;

    // Check that there is enough space to do this
    if (heapstart + allocsize > heapend)
        return NULL;

    heapstart = (unsigned char *)
      (((unsigned int) heapstart + 3) & 0xfffffffc);
    result = heapstart;
    heapstart += allocsize;
```

```
    heapnext = (unsigned char *)
      (((unsigned int) heapstart + 3) &
          0xfffffffc);

    return result;

}
/*
    Reset our quick and dirty memory manager.
    This function deallocates all non-perma
    nent arenas.
*/
void mmgr_Reset(void)
{
    heapnext  = (unsigned char *)
       (((unsigned int) heapstart + 3) &
          0xfffffffc);
}
/*
    Allocate a memory block. Returns pointer
    or NULL if there isn't enough heap left.
*/
void *mmgr_Alloc(size_t allocsize)
{
    void *p = (void *) heapnext;

    heapnext += (unsigned int) allocsize;
    heapnext = (unsigned char *)
       (((unsigned int) heapnext + 3) &
          0xfffffffc);
if (heapnext > heapend)
           return NULL;
    return p;
}
```

At the start of your program, you would call
mmgr_Initialize(). Each time you enter a piece of code that requires a temporary heap, you would call mmgr_Reset(). The
mmgr_Alloc() function is designed to resemble malloc; you can
convert existing code to use this basic memory manager simply
by defining a macro like this:

```
#define malloc(x) mmgr_Alloc(x)
```

The mmgr_PermAlloc() call allocates a space in the heap, and reduces the heap size to make the allocated area off-limits and unaffected by mmgr_Reset(). Only space left unused in the heap after calling mmgr_PermAlloc() is available for transitory allocation via mmgr_Alloc().

This function is provided as a convenience feature, allowing you to perform runtime balancing of memory usage between different functions. For example, if your application involves storing a list of username/password pairs and a list of filenames, you could use mmgr_PermAlloc() (at boot time) to allocate space for the username/password list and the filename list. If the user runs out of space in either list, you can provide a user interface element that allows him or her to adjust the space allocated to each of these functions, and store this preference in some non-volatile device attached to your system. This is at least slightly more elegant than having hardcoded size limits on each list.

Note that the code above always aligns allocation requests on a 4-byte boundary, since some 32-bit platforms have restrictions on dealing with non-aligned data. Also be aware that this particular example is non-reentrant. Unlike a more general-purpose memory manager, the function of this code is not to arbitrate between multiple users of the heap and cater to them all as far as possible. Rather, it allows a single user to keep track of arbitrarily sized blocks of memory, typically during a single call to a complex function.

All of the above types of data storage issues and conflicting requirements are illustrated rather nicely in a consumer electronics product I designed for my employer in 1999. This project was a digital picture frame with a color LCD screen and slots for CompactFlash and SmartMedia flash memory cards. The normal usage pattern for this device is that the owner will take some photos with his or her digital camera, then put the card in the picture frame to browse the pictures. It's necessary for the picture frame to build (in memory) a list of all the files on the inserted card(s), along with information about rotation, color adjustment and other special effects the user might want to apply to each image. Because the user wants to be able to control which pictures he or she will see in the slideshow, we need to be able to add and subtract individual items from the list in memory. We want to be able to store the longest possible slideshow in memory; perhaps several thousand entries, since modern removable flash media are dense enough to store a large number of images. For this purpose, I implemented a packed string list as described previously. (In fact, it was slightly more complicated than the example above, because I had to store a little binary data along with the string—but the principle is the same.)

The picture frame also had to contain codecs for various image formats, and most of these codecs have varying memory requirements depending on the type of input data. (For example, memory requirements are very different for decoding sequential versus progressive JPEG files, even though both types are handled by the same codec. An even better example is TIFF, which has many different sub-formats that need to be handled differently.) Since the device I was building is single-tasking, there is no situation where we could simultaneously need to be decoding (say) a JPEG and a TIFF. Thus, I could safely reserve a single RAM area for image decoding scratch space and allow each codec to assume it had unlimited access to that area. To achieve this, I implemented a simple memory manager of the type described above.

Rigorously avoid writing recursive functions. On your super-powered demonstration platform, blowing the stack out to 300K might not be a problem, but on your real hardware you probably won't be able to afford so much space. Worse still, the nature of recursive functions is that you probably don't know at compile time just how deep the recursion is going to get. The archetypical example of such a function is scanning directories on a filesystem; you have no idea how deeply someone might have nested directories on the storage medium, and you *must* have some checking to avoid runaway conditions. It is usually very difficult to accurately estimate worst-case resource usage (taking into account asynchronous events such as interrupts, and contention for resources such as drivers for storage media). It's much easier to design your code intentionally to minimize the effects of these issues.

If it's absolutely necessary to write recursive functions, then you should explicitly check stack usage each time you descend a layer of recursion. You should provide a failure mechanism whereby attempting to go too deep will either fail the entire parent call with an error, or otherwise communicate to the user that the input data structure is too complex. To stave off this error condition as long as possible, make sure that your function allocates the barest minimum of local variables. You should be even more careful not to overuse any system-global resource (such as file descriptors or network sockets) in your recursive function. If possible, you should save the state of such resources internally and deallocate the real handle, socket or other structure before calling down to a deeper level of recursion.

Keep in mind that it's much better to have a function that "bottoms out" with an error message when presented with certain inputs, than to have a function that will randomly corrupt some memory areas if someone gives it unexpected input. Unless you put in the extra effort to include sanity-checking, recursive code is liable to be deceptively robust on your demonstration platform, but unreliable on the real hardware. It's considerably easier to add the requisite checks and robust design methodologies when you're first writing the code than while you're porting it to a more constrained system.

If feasible, avoid the use of multitasking features such as threads and subprocesses. There are a multitude of reasons for this. The first is simply that by using threads, you are assuming that your final operating system will have some kind of preemptive multitasking capability. Although many embedded operating systems do have this feature, there are plenty of occasions where the additional system overhead is unnecessary. Additionally, there are several different "standard" threading APIs, as well as many proprietary systems, and even a single operating system may have more than one way of starting a secondary task from a main program, with correspondingly different limitations and features according to the method chosen.

Secondly, having multiple pieces of code running "concurrently" (from a logical standpoint, at any rate) opens up a new dimension of complexity in debugging, to say the least. It's much easier to serialize access to resources if the program flows in a linear fashion. Not all debuggers can properly handle multithreaded code, and even if yours can, running multiple tasks can open you up to a variety of interesting and hard-to-debug synchronization problems. Race conditions and similar problems might not be obvious on your demonstration platform, and can be exceedingly difficult to debug once you move to the real hardware.

Note that I'm certainly not saying that threaded code is inherently evil; if your program can be simplified by using threads while still remaining robust, by all means go ahead and use them. I am simply advising that you first make a conservative assessment of the downsides, and preferably don't start developing multitasking code until you've chosen an operating system for the final product.

There is one very useful application of threads (or other multitasking technology) on a demonstration platform, and that is to emulate asynchronous interrupt-based events that can occur on the real hardware. For example, the PC-based prototype system I mentioned in my horror story at the beginning of this chapter had a set of pushbuttons and indicators driven from the PC's parallel port. Our real hardware design ran these buttons to I/O ports with interrupt capability so that we could read these controls asychronously. However, most of the lines on the PC parallel port can't generate interrupts, so it was necessary to poll these controls on the prototype. In order to make the program flow work similarly to the way it would work on the real hardware, we ran a separate process to poll and debounce the controls at a moderate rate (approximately 50Hz). This process updated a shared memory area representing the control state. In the real program, we simply replaced the slave process with an interrupt handler and the overall behavior of the ported program was exactly the same as on the demonstration hardware.

If you are writing a proprietary function to accomplish a task that would normally be provided by an operating system, try to make this function's top-level interfaces similar to those normally found in other operating systems. For instance, if you are writing a filesystem, your life will be simplified if you emulate the standard streams calls such as fopen(*filename*, *access-mode*), fclose(*file-pointer*), and so on. It is usually not much work to dummy out unused parameters or return default (or error) values for unsupported functionality, and it makes excellent sense to use a well-known interface layer to insulate your program's main functionality as far as possible from operating-system-specific code.

If you're not sure of what best to emulate at the top layer, then consider POSIX compatibility, at least in terms of the parameters and function names, and you won't go far wrong.

Maintain a division between platform-dependent and platform-independent code. This is simply sound firmware engineering practice. One good starting point, assuming that you're developing with gcc and newlib, is to make sure that your program accesses resources (devices, files and so on) only via industry-standard APIs provided within newlib. That way, all you have to do in order to move to a new hardware platform is to port the required portions of newlib and relink your application against the new library. (You will also be able to run your application on top of many different operating systems with relatively little platform-specific handling, since the APIs in newlib are a subset of those provided on almost every major desktop operating system.)

There are, however, a lot of functions that aren't provided by newlib and hence can't be abstracted through this method. One of the most common sets of such functions that 32-bit system developers tend to want is a graphical user interface library. Depending on your application, you might just want simple functions to output text onto a bitmapped display device, all the way up to a massively complicated windowed operating environment with exotic 3D graphics primitives, YUV overlays for motion video playback, and so on.

Exactly how you choose to solve your GUI needs involves trading off simplicity of design, easy portability and total development effort. There are a large number of ready-made projects intended to help the embedded developer with this potentially large development task—probably the two most popular of which are wxWindows and nano-X. At the extreme high end are fully-functional GUI environments like Xwindows, which were not originally intended for embedded applications but are finding increasing use in "quasi-embedded" devices built around PC-compatible single-board computers and other relatively powerful hardware platforms.

In general, attempting to write your own GUI from scratch means reinventing many wheels; it's almost always a much better cost/performance plan to adapt someone else's free code than to develop your own. (It might not seem like that when you first approach the task, though, because you'll be at the bottom of the learning curve for the library you've chosen to work with.) If you're determined to write your own system, then first be sure that you can really justify this amount of work. This involves working out *exactly* what your requirements are, and sketching out a hardware abstraction model before you start writing a line of code.

For example, I regularly work on the firmware and hardware for a family of digital imaging devices. These devices range from simple low-end consumer units with quarter-VGA sized, 12bpp color LCD screens up to very large devices (intended for the commercial market) with enormous SXGA-resolution 24bpp color screens, among many other features. It's obviously desirable for us to maintain as much code commonality as possible amongst the various models.

Because we have very specialized, well-understood needs that aren't completely addressed by any of the common GUI options, I felt it was worthwhile to develop our own GUI. Despite the varying color depths of our various products, I chose to have as much as possible of our code work in an RGB 8:8:8 colorspace, because this allows almost all of the program to be hardware-independent—all the code knows is that it's manipulating a 24bpp internal virtual image buffer of known dimensions, and the process that translates this buffer into the framebuffer hardware's native format is invisible to almost all of the program. Obviously, this has some performance drawbacks, because on most hardware platforms, a fair amount of redundant data is being manipulated, but it is an acceptable tradeoff for our purposes. (12bpp is a particularly annoying color format to handle, because a single pixel does not occupy an integral number of bytes. Despite being the lowest-bandwidth direct color format we support, it is also far and away the lowest-performance because of this issue.) Moreover, having a single format for internal image representation has allowed us to spend a lot more time developing a rich set of top-level APIs and optimizing these APIs. Each time we develop a new hardware platform, all that requires porting is a single call that renders a logical framebuffer (at 24bpp) onto the physical framebuffer at its native color depth and format. Because the high-level APIs are code-identical across platforms, we can be more confident that they will exhibit consistent behavior; and because of the relatively small code volume, debugging and optimization is facilitated.

For Web-downloadable demonstration purposes, we have even written DirectX "wrappers" that allow our real embedded application code to run directly as a Windows application. This has obvious benefits as a sales demonstration tool. One further step we have discussed without implementing, though it is certainly technically possible, is to port our GUI layer down to work as an ActiveX control so that we can demonstrate our embedded application live on a web site.

Contrast this happy scenario to the alternative: If we had decided to put the hardware abstraction cut-off point a little higher, we could have implemented the same standard set of APIs for functions such as opening a window, printing text, anti-aliasing, etc. However, these APIs would all work directly in the native framebuffer format. While it would undoubtedly result in faster code (at least on those platforms that operate at color depths less than 24bpp), it would mean that each time we build some new piece of hardware that has its framebuffer memory laid out slightly differently, we have to rewrite *every* API rather than a single rendering API. Moreover, each time we want to add a single feature or optimization, there is a huge amount of work involved to backport the new code to all possible target variants.

The above is one possible illustration of the type of abstraction that it is useful to consider in your application. Another subsystem that benefits greatly from careful layering is a filesystem. I have maintained several independent projects that implemented their own DOS-compatible filesystem (already extant at the time I inherited the projects in question). None of these filesystem modules was adequately abstracted, and I am quite sure that the initial debugging process for all of them was unnecessarily painful. Porting them to any new platform would have been even more painful.

At its upper, application-interface layer, even a minimal filesystem consists of APIs that work with filenames and streams of data bytes provided by the application, and that return logical file descriptors and streams of data bytes off the underlying storage medium. If we consider a desktop PC as a maximal example, it is clear that a single conventional hierarchy for filenames and path specifications can refer to files stored and retrieved in widely disparate ways. There are network filesystems referring to remote files, ISO9660 or UDF filesystems referring to CD-ROM and DVD-ROM media, and various filesystems typically used on random-access media, such as the ubiquitous FAT, Linux ext2, and so on.

Therefore, the first natural layering zone in the hierarchy of a well-designed filesystem is at the point where filenames are handed off to specific filesystems. Above this point, all code can be generic. A file descriptor, for instance, is just a pointer to some data structure. This structure presumably identifies which filesystem driver handles requests for the file in question, but manipulating those structures doesn't directly require knowledge of the underlying filesystems.

Continuing to use the PC as an illustrative example, it's clear that we can also have a single filesystem in use across different storage media types. For instance, we can use the same FAT16 code on an IDE hard drive, a floppy disk, a PCMCIA card, a SmartMedia® flash memory card, a SCSI-connected Iomega Zip® drive, and so on. All the FAT16 filesystem code cares about is reading and writing 512-byte sectors; it doesn't need to know anything about the minutiae of how bytes are sent to and from the storage media. Clearly, therefore, the next level in the abstraction layer should be to separate the filesystems from the low-level device drivers that perform sector-level I/O and other hardware-specific tasks such as verifying the presence of a storage medium in drives that support removable media.

By doing this, we can break the debugging effort into easily manageable chunks—the FAT16 code, for instance, can be debugged by dumping a small known-good FAT filesystem from a hard drive partition into a file, and testing our FAT16 filesystem over a dummy storage device that reads and writes the disk image file, using code similar to that below:

```c
/*
    Code to read a single physical sector
    number from a disk image
*/
int ReadImageSector(char *buffer, int
sector_number)
{
    FILE *image_file = fopen("disk.image", "r");
    int result;

    if (!image_file)
        return 0;
    fseek(image_file, sector_number*512,
    SEEK_SET);
    result = fread(buffer, 512, 1, image_file);
    return result;
}
/*
    Code to write a single physical sector
    number to a disk image
*/
int WriteImageSector(char *buffer, int
sector_number)
{
    FILE *image_file = fopen("disk.image", "w");
    int result;

    if (!image_file)
        return 0;
    fseek(image_file, sector_number*512,
    SEEK_SET);
    result = fwrite(buffer, 512, 1, image_file);
    return result;
}
```

Similarly, we can develop our low-level device drivers very quickly because there is no interdependency with an overlying filesystem driver. All we need to do is write sector read and write functions, along with whatever additional functionality might be required to enumerate sub-devices, check media state, operate an eject mechanism, and so on. Given that our filesystem code has proven itself in vitro while running on top of dummy image-based functions like those above, we can have a high degree of confidence that the entire filesystem hierarchy will work correctly when the various modules are brought together. Conversely, we can be reasonably confident (hardware interactions aside) that when we need to add support for some hitherto unknown storage device, we will be able to do so simply by adding a new low-level device driver.

You should notice that the recurring theme in my text above is to **develop code that can be debugged anywhere**, even if it can't truly be said to "run" everywhere, and that can be divided into **modules that can be reliably debugged in an incomplete system**. If you make these goals your mantra, you will be able to get the bulk of your program debugged on your desktop PC, where loading a new version takes no more effort than compiling it and hitting a "run" button. You can confine the tedious iterative compile-upload-flash-test process to debugging a very small volume of platform-dependent code, thereby speeding you towards a reliable, efficient final product.

## Chapter 6

Useful Vendors and Other Web Resources

This section contains a short directory of vendors and products I have found helpful in my efforts to develop products within a constrained budget.

Advantech *(www.advantech.com)*

Advantech is a manufacturer of (among other things) x86-based PC-compatible single-board computers. Although there are many other such manufacturers, Advantech has two advantages; firstly, they target non-industrial applications with a range of reasonably priced SBCs, and secondly you can buy many of their products online directly. Most of Advantech's competitors concentrate on boards with extended temperature tolerance, ESD hardness or other specialized industrial requirements, which make those boards far too expensive for the average reader of this book.

Atmel *(www.atmel.com)*

At the time of writing, Atmel's AT91 series of ARM-based microcontrollers is one of the cheapest routes to a new ARM-based design. The microcontrollers themselves are cheap, and the evaluation boards are an order of magnitude cheaper than the boards from most other ARM vendors.

The ARM Linux Project *(www.arm.linux.org.uk)*

Your central starting point for resources related to running Linux on ARM-cored microcontrollers.

Note: If you intend to run Linux on a microcontroller with no memory-management unit, you should visit the ucLinux site (www.uclinux.org) instead.

The GNU Project *(www.gnu.org)*

This site is your jumping-off point for documentation and the latest versions of the GNU development tools. Note that newlib, though integrated with the GNU toolchain we discuss in this book, is *not* a GNU project; it is maintained by Red Hat (see below).

Linux Devices *(www.linuxdevices.com)*

This is a useful portal to news and vendor directories of Linux-based embedded systems, particularly off-the-shelf consumer devices and development tools.

Macraigor Systems LLC *(www.macraigor.com)*

Macraigor sells a range of JTAG/ICE debugger modules for various processors including ARM, Motorola CPU32, MIPS, PowerPC and XScale. Their entry-level product, Wiggler, is one of the lowest-cost commercial hardware debugging products available for these high-end microcontrollers.

Monta Vista Software, Inc *(www.hardhatlinux.com)*

These people are the developers of Hard Hat Linux, one of the best-known realtime adaptations of Linux. This is commercial software, not available for unrestricted download.

Opencores *(www.opencores.org)*

Opencores is the central reference on the web for free open-source HDL core IP to integrate into your FPGA-based project. Free cores available at the time of writing include LCD/VGA controllers, memory controllers, microcontrollers, communications devices and cryptographic functions.

Red Hat, Inc *(www.redhat.com)*

In addition to their well-known Linux distribution, Red Hat is the maintainer of the free eCos embedded operating system. You can find the eCos homepage at sources.redhat.com/ecos. Red Hat also maintains the newlib project at sources.redhat.com/newlib and the Cygwin project at sources.redhat.com/cygwin.

Trenz electronic GmbH *(www.trenz-electronic.de)*

Trenz electronic sells, among other things, low-cost FPGA evaluation boards based around Xilinx parts. If you are considering a design built around an FPGA, Trenz's boards are well worth considering.

Xeltek, Inc. *(www.xeltek.com)*

Xeltek sells a wide range of device programmers, at comparatively inexpensive prices. In particular, their SuperPro® Z programmer is one of the best such devices I have seen (on a price vs. supported devices basis).

Index of CD-ROM Contents

The CD-ROM included with this book contains various useful pieces of software referenced in the text. Below is a brief index of these software items, with their locations on the CD-ROM and installation instructions. This disk was mastered with Microsoft's Joliet extensions for long file names; if your operating system does not support Joliet, then you may see truncated ISO9660 filenames in some cases. In the case of the example sourcecode files, you will usually have to translate any mutilated filenames back to their original "long," lower-case name before the build process will work properly.

Cygwin (Version 1.3.16-1)

Location: cygwin\setup.exe

Download From: http://sources.redhat.com/cygwin/

To install, double-click the setup.exe file

EAGLE (Windows version 4.09r2, English)

Location: eagle\eagle-4.09r2e.exe

Download From: http://www.cadsoftusa.com/

This is a single-file installer; simply run the executable to install the product. If you are using Windows NT, 2000 or XP, after installing the software you should go to the Options menu, select "User interface", and in the dialog that appears, ensure that the "Always vector font" box is checked.

EAGLE (Linux version 4.09r2, English)

Location: eagle/eagle-4.09r2e.tgz

Download From: http://www.cadsoftusa.com/

To install, decompress the tarball with tar zxvf eagle-4.09r2e.tgz. Change to the eagle-4.09r2 directory thus produced, and run the script ./install to begin an automated install process.

Example sourcecode files

Location: sourcecode/

Download From: http://www.zws.com/

Instructions on how to compile and load these sample programs are provided in the chapter headed "Example Firmware Walkthroughs and Debugging Techniques".

GNU binutils (Version 2.13.1)

Location: gcctools/binutils-2.13.1.tar.gz

Download From: ftp://ftp.gnu.org/gnu/binutils/

Installation instructions for this product are provided in the chapter headed "The GNU Toolchain".

GNU gcc (Version 3.2)

Location: gcctools/gcc-3.2.tar.gz

Download From: ftp://ftp.gnu.org/gnu/gcc/

Installation instructions for this product are provided in the chapter headed "The GNU Toolchain".

GNU gdb (Version 5.2)

Location: gcctools/gdb-5.2.tar.gz

Download From: ftp://ftp.gnu.org/gnu/gdb/

Installation instructions for this product are provided in the chapter headed "The GNU Toolchain".

Newlib (Version 1.10.0)

Location: gcctools/newlib-1.10.0.tar.gz

Download From: http://sources.redhat.com/newlib/

Installation instructions for this product are provided in the chapter headed "The GNU Toolchain".

About the Author

Lewin Edwards was born in Adelaide, Australia and currently works for Digi-Frame Inc. in Port Chester, NY. He has worked for over four years in security and encryption applications, and he has dabbled in engineering speaking and musical toys. For almost five years he has been implementing digital imaging and multimedia applications on a variety of embedded platforms based around PA-RISC, ARM, Intel and other 32-bit processors. He was most recently published in Embedded Systems Programming magazine and can frequently be found on Usenet in comp.arch.embedded. For relaxation, he works on simpler 8-bit realtime control applications such as an ongoing model submarine project.

Index

Numbers and symbols
680x0 series, 14

A
Arm, advantages, 18-19
arm-linux, 74
Atmel EB40, 29-30, 153-157
    hardware bug, 156
    memory layout, 155-157

B
BGA, 59
blind debugging, 194-198
board layout techniques, 56
bootstrap ROM, 23
breakpoints, 149-150
building the toolchain, 71

C
C++, special considerations, 136
CD-ROM index, 223
CFLAGS, 85
comments, 88
conditional assembly directives,
    108-111
core, choosing the correct one, 13-19
core, debugging, 13
CVS, 40

Cygwin environment, 51, 71-73, 223

D
datasheets, 11
debugging, 81, 139, 213
    blind, 194-198
development board, 11
development cycle, 21-22
development hardware, choosing,
    21-29
device drivers, low-level, 219
discrete cosine transform (DCT), 20
dual-boot systems, 51
dynamic memory allocation, 205

E
eCos operating system, 40
embedded operating systems, free,
    36-44
ENTRY command, 118

F
filesystem, 217-219
flash-loader program, 173-180
FPGAs, building custom peripherals
with, 19-21
FreeDOS, 35
free software, 45-50

G
gas (GNU assembler), 87-88
gcc, 79, 215
    location on CD-ROM, 224
gdb (GNU debugger), 143-145
    command-line options, 145-146
    location on CD-ROM, 225
General Public License, see GPL
glue code, 199
GNU make, 80
GNU build environment, overview,
    76-77
GNU toolchain, 12, 35, 71-151
GPL, 45-48
graphical user interface (GUI),
    215-217

H
hardware design language (HDL), 20
header files, 86

I
i386-pe, 74
Id, 87
Ida Pro, 27
image formats, codecs for, 212
info pages, 79
initialization, 170
input section descriptions, 132-134
intellectual property, 50
internationalizing a product, 207
internet resources, 221-225
interrupts, 199-201

L
laboratory equipment, 30
Ld, GNU linker, 114-117, 137
    built-in functions, 123-124

LED flasher program, 158-172
Linux, 37-38
    limitations of, 37
    Hard Hat Linux, 37
local memory address (LMA), 116

M
m68k-coff, 74
m68k-elf, 74
Macraigor Wiggler, 26, 35-36, 51, 76
macros, 111
make, 79
makefiles, 80
master board, 58
memory
memory manager code, 208-211
microcontroller, selecting, 9-13
MinGW, 52
MIPS, 18
mips-elf, 74
Monta Vista Software, 37
multimeter, 30-31
multitasking features, avoiding, 213

N
named memory regions, 134-135
Newcom Webpal, 25-26
newlib, 75, 215
    location on CD-ROM, 225
NetBSD, 39, 48

O
objcopy, 138
objdump, 139
OBJS, 84
Opencores, 21
open-source software, 44
OpenWatcom, 34

operating systems, 32
choosing, 51-53
free, 36-44
oscilloscope, 31
add-ons for PCs, 32
output section descriptions,
124-125, 131
overlay section descriptions, 127-130

P
Palm OS, 41
patents, software, 50
PCB CAD software, 67
PCB layout, 28, 53-59
software, choosing, 65-68
vendor, 57
PCB prototyping, 66
PDA, 14, 22
platform-dependent code, 215
portability techniques, 203-125
portable code, 199
POSIX compatibility, 215
prototyping, 203
pseudo-operations, gas, 96-108

R
ready-made operating system, 43
real-time operating system (RTOS),
37
recursive functions, avoiding,
212-213
releasing your product, 46
reliability considerations, 213-220
repurposing, 24
resources, internet, 221-225
ROM-startup program, 180-194
RTLinux, 38

S
section directives, 90-96
SECTIONS command, 118-119
sections, code, 90
shareware software packages, 65
sh-elf, 74
size utility, 143
soldering iron, 31
source code, location on CD-ROM,
224
startup code, 180-194
SuperH, 18
surface-mount chips, 59
surface-mounting by hand, 62-65
symbol assignments, 119
symbols and labels, 88-90
system on chip, 19

T
target memory, examining, 148
threads, 213-214
emulating asynchronous
interrupt-based events, 214
toolchain, choosing, 32
free development toolchains,
32-36
Trenz electronic, 20

V
vendor list, 221
Virgin Webplayer, 26
virtual memory address (VMA), 116

W
watchpoints, 150
Wiggler, Macraigor, 26
Windows CE, 41

X

x86 family, advantages, 14-17
xscale-elf, 74